WATER RESOURCES DEV

IN DEVELOPING COUNTRIES

WATER RESOURCES DEVELOPMENT IN DEVELOPING COUNTRIES

DAVID STEPHENSON

Department of Civil Engineering, University of the Witwatersrand, Johannesburg, South Africa

MARGARET S. PETERSEN

Department of Civil Engineering, University of Arizona, Tucson, AZ 85721, U.S.A.

ELSEVIER

Amsterdam — Oxford — New York — Tokyo   1991

ELSEVIER SCIENCE PUBLISHERS B.V.
Sara Burgerhartstraat 25
P.O. Box 211, 1000 AE Amsterdam, The Netherlands

Distributors for the United States and Canada:

ELSEVIER SCIENCE PUBLISHING COMPANY INC.
655, Avenue of the Americas
New York, NY 10010, U.S.A.

ISBN 0-444-88956-6

PREFACE

Water resources exploitation has been regarded as a way of initiating economic development in many countries. It was successful in creating employment in the western U.S.A. in the 1930's and in subsequent settlement and development of the area. It was successful in providing food and infrastructure in India following British colonization in the 19th century. Other cases have risen where the projects have been of limited use in Africa, or have been criticised for lack of lateral vision. Environmentalists and sociologists have criticised engineers for constructing concrete monuments with insufficient attention to the background.

Planning ideas have changed. Thorough environmental studies, sociological and economic studies now preceed project formulation. Justification solely on the basis of benefit cost studies is no longer sufficient for many development agencies. The broader approach is introduced in this book, but the real emphasis is on the situation in and needs of developing countries. These countries are rarely self-sufficient and as a result are dependent on aid for development. Assuming they do wish to develop, and the rest of the world wants them to develop, some impetus is needed. Private enterprise can be encouraged to establish in these countries, but then finance and resources are required. Large scale public works is still generally regarded as the best way of developing infrastructure and employment to create stable labour forces and political scenarios. Water is often the most accessible resource for exploitation in undeveloped countries. Irrigation works provide food, and hydro power is generally maintenance-free, both suitable for third world areas.

Developing countries differ from developed countries in many ways, requiring different approaches in planning water resources development. One difference is the lack of basic data in underdeveloped or developing countries. There may be no established hydrological network or even rain gauge network. Flow records are therefore lacking, or at least are of insufficient duration and accuracy. Sophisticated simulations are, therefore, not warranted and cannot be done without considerable guesswork. Simplistic storage – draft relations may be the best that can be done.

Economic data may also be poor and unrepresentative. Prices are distorted and may not represent true worth to the country. Shadow values may be more applicable and these could be generated to enable a master national plan to be achieved. Planning methods, therefore, have to be developed to suit the circumstances.

Some of the problems experienced in developing water resources in developing countries are described in this book, and methods of solution based on the limited experience of the authors, are offered. These range from use of unbiased common sense, coupled with a close understanding of people's requirements, to a comprehensive computer simulated planning model.

Some types of water resources development are described in more detail. These include irrigation, hydro electric power and rural water supply. Sections on socio-economics and human resource development are also included, as well as on data collection, and project planning.

Lessons from the failure of multimillion dollar projects are not hard to come by, and examples and pointers which will assist future planners are given. Attention is paid to the need for aid to include training and to stimulate local economies. However big water projects appear, they cannot escape the effects of the rest of the country's economy. Attention is also drawn to environmental problems, particularly soil erosion, often caused by water resources development. The fact that water resources development cannot be carried out by engineers only, is recognized. The input of many professions, and vast experience, is needed.

The authors have drawn on international case studies. Much of the material has been presented in postgraduate courses by the authors. Margaret Petersen obtained a sound basis in applied engineering with the U.S. Army, Corps of Engineers in the field of hydraulic engineering and water resources planning before commencing teaching in the Department of Civil Engineering at the University of Arizona. She has a long-standing interest in water problems in lesser developed countries and has lectured in the Far East and Africa. David Stephenson obtained extensive experience in consulting in Southern Africa before entering the academic world. He has worked amongst developing communities planning and designing water projects and established training programmes. (He was kidnapped by Renamo rebels whilst on a hydrological assignment in Mocambique and escaped after months in captivity). Robert Clark contributed much of Chapter 8; he was Director, Office of Hydrology, U.S. National Weather Service and presently is in the Department of Hydrology at the University of Arizona. He has had extensive international experience, especially in China. Brian Hollingworth of the Development Bank of Southern Africa provided much of Chapter 3, Tony Venn, of Loxton Venn, Chapter 9 and many other contributed in various ways. The manuscript was typed and arranged by Janet Robertson.

CONTENTS

CHAPTER 1. WATER RESOURCES PLANNING OBJECTIVES

CHAPTER 2. SOCIO-ECONOMIC FACTORS

CHAPTER 3. ECONOMIC PRINCIPLES

CHAPTER 4.   SYSTEMS ANALYSIS AND OPTIMIZATION

CHAPTER 5.   DECOMPOSITION OF COMPLEX SYSTEMS

CHAPTER 6.   A PLANNING MODEL

CHAPTER 7.   RESERVOIR SIZING

CHAPTER 8.   HYDROMETEOROLOGICAL NETWORK DESIGN
AND DATA COLLECTION

CHAPTER 9.   SOIL EROSION AND SEDIMENTATION

CHAPTER 10.   IRRIGATION

CHAPTER 11.   RURAL WATER SUPPLIES

CHAPTER 12.   HYDRO ELECTRIC POWER DEVELOPMENT

CHAPTER 13. HUMAN RESOURCES

CHAPTER 14. ENVIRONMENTAL AND SOCIAL IMPACT ASSESSMENT

CHAPTER 1

WATER RESOURCES PLANNING OBJECTIVES

"Water resource planning and development" is a broad term having numerous meanings and is defined in different ways by different groups or individuals. Further, planning studies are of varying scopes and different programs would involve different measures for studies:

- At the national, regional (or river basin), or local level.
- To meet long-range or short-range objectives.
- To meet single or multiple needs and problems.
- For urban or rural areas.
- For humid or arid and semi-arid areas.
- Using surface water sources only, ground water only, or both conjunctively.
- In developing countries or industrialized countries.
- For a river basin entirely within one country or an international river basin in several countries.

In water resources planning for developing countries one is faced with most of these varied facets of the overall problem.

Additionally, there are three broad philosophical bases for establishing objectives for resource development in developing areas:

- One is simply to improve the quality of life and well-being of the people by providing goods and services (e.g. safe water supply, irrigation water, improved public health) and training and education programs. This objective is recognized by the United Nations Water Decade.

- Another involves improving the quality of life and social well-being by modifying life styles in various ways (e.g. sedentarization of nomadic people, changing farming methods, establishing new population centres, etc.)

- A third is by investing money which gets into circulation, providing jobs and training to stimulate the economy and increase the rate of economic advancement towards independence of first world aid.

2

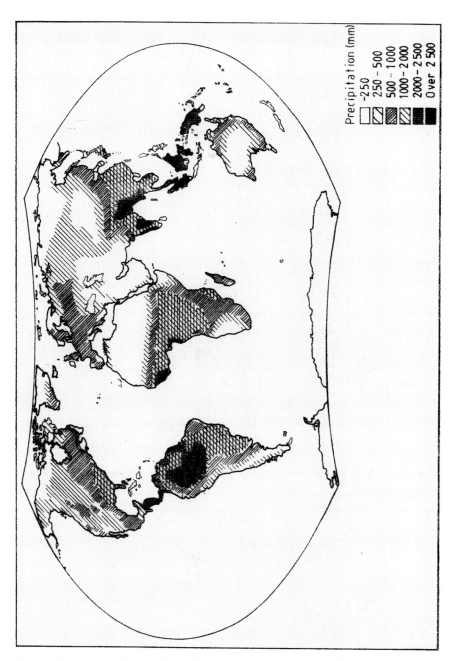

Fig. 1.1   Annual World—Wide Precipitation
            (Source: U.S. Department of Commerce, Environmental Science
            Services Administration, 1969).

3

Fig. 1.2   Annual World-Wide Evaporation (Lvovitch, 1973)

Fig. 1.3   Annual World-Wide Total River Runoff in mm.

Development of water resources is an increasingly important factor in economic and social development of societies everywhere. Water scarcity can limit national and social development of societies everywhere, while uncontrolled surplus water (flooding) can inflict severe economic and social losses. Primary responsibiity for development and control of water resources usually rests with national governments for several reasons: (1) the crucial importance of water (2) the complex institutional, financial and legal considerations usually involved; and (3) because political boundaries do not generally follow drainage basin boundaries.

The quantity and quality of water available to people everywhere is one of the primary factors influencing quality of life, through food production, flood damage, navigation, public health, and energy production. To manage water effectively on a national scale is to control it for the benefit of all people as it passes through the natural hydrologic cycle from precipitation to runoff or infiltration, and then to the oceans. Effective management is the efficient (least cost) and timely implementation of the right measure (including policies, regulations, construction of projects, and operation) in a way that is socially acceptable and environmentally sound.

WATER RESOURCES

Development and management of water resources involve modification of the hydrologic cycle to regulate the natural water supply to better meet human needs. Planners must recognize the close interrelationship of the hydrologic cycle with other systems, including:

- Land use, soil conservation, and watershed management.
- Groundwater supply and use.
- Drainage and aquatic weed control.
- Demographics (population characteristics and distribution).
- Economics.
- Social well-being.
- Flora and fauna.
- Public health and control of disease vectors.

Water on the planet earth is not equally distributed or equally available everywhere. Of the order of 97 percent of all water on earth is too salty for human use, and only 3 percent is fresh. Of what fresh water there is (see Fig 1.1 - 1.3), about 75 percent, is locked in the polar

6

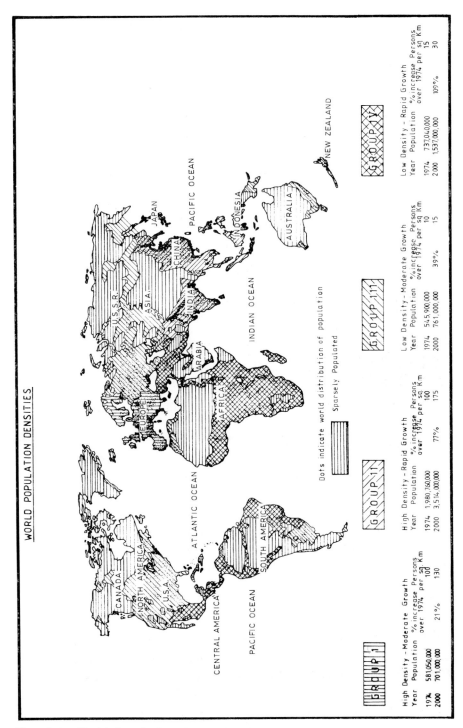

Fig. 1.4   World Population Densities

icecaps and glaciers and is generally unavailable. Of the remaining 25 percent available for use (from rivers, lakes, and ground-water basins), only about 1.2 percent of the volume at a given time is in the form of surface water, while 98.8 percent is ground water. However, when we consider volume available over a time period, the percentages change because surface water flows at significant velocities while ground-water movement is very slow (van der Leeden, 1975).

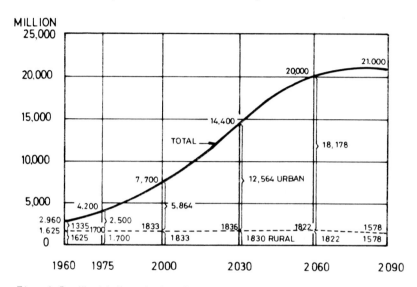

Fig. 1.5   World Population Projection
(Source: Doxiadis, Water for Peace, 1967)

It is essential to consider all ground-water supplies when planning water development and management to meet long-range needs because they are a major resource. Also, shallow ground-water deposits are interconnected to surface-water supplies and both are vulnerable to short-term drought effects, while deep ground-water supplies are relatively unaffected by drought.

NATIONAL POLICY

For best utilization of available resources, coordinated development and management programs are required at both the national and local levels. National governments must consider the interests of future generations as well as those of the present population in their operations and, therefore, must work with very long time frames. To establish a national policy for wise long-term use of surface and ground water involves:

1. Establishing both long-range national objectives (such as food self-sufficiency, meeting water needs with adequate supplies of acceptable quality over time, and improving social and economic conditions) and compatible short-term goals for individual river basins and local areas (such as providing water to bring additional lands into cultivation, providing safe drinking water for villages, reducing flood damage, and meeting growing needs for municipal and industrial water and electricity).

2. Establishing priorities and programs for long-range and near-future development and operation, recognizing financial budget constraints.

3. Scheduling development through staged construction.

Water Resources Planning

Planning is a process, a systematic way of investigating a problem. It is both an art and a science, an exercise in managing information, acquiring information, evaluating it, and analyzing it and then making decisions. It is a systematic way of establishing long-range national objectives and devising long-term strategies and short-term action programs to achieve those objectives and to implement the strategies by solving present and projected future problems. Planning is a systematic study of alternative solutions to a problem or need, including the costs, benefits, and adverse impacts of the alternative, and selection of the best plan.

For planning to be successful, the plans developed must be technically and politically implementable, affordable, socially acceptable, and environmentally sound. Planning _per se_ is worthless. Planning has value only when plans are implemented successfully and people receive the benefits projected. It is essential that all policies, institutional arrangements, and programs be appropriate for the local culture and that they be understood and accepted by all those directly affected.

The need for long-range planning has become more critical in recent years with population growth and increased development and utilization of the world's resources. There is no substitute for water, and the objectives of water resource planning centre on wise use of resources to avoid future shortages that might otherwise limit a nation's economy or the social well-being of its people.

Because manpower and funding are limited, priorities for resource development and utilization must be established through systematic planning. Wise planning and integrated development in a river basin serve to assure that no isolated irreversible plan for part of the basin is implemented that might limit future freedom of choice and also that costs of water resources control measures are minimized.

LEVELS OF WATER RESOURCE PLANNING

We identify three "levels" of planning, associated with three areas of different geographic extent: national; regional or river basin; and local areas. Planning for these different areas is conducted to different degrees of detail and for different purpose.

Level A. – Level A studies ("framework" studies) are substantially inventories of resources and of problems and needs. They are of national scope and are considered with long-range (up to 50-year) projections for large geographic areas encompassing two or more river basins. Based on current demographic, economic, social, and environmental indicators, forecasts are made of future trends, problems, and needs related to water resources. Framework studies are designed to:

1. Inventorize the extent of water and related land problems, needs, and desires of people for the development, utilization, and conservation of water and land resources.

2. Indicate general approaches that appear appropriate for solution of the identified problems and needs.

3. Identify specific geographic areas with complex problems where more detailed regional or river basin (Level B) or implementation (Level C) studies are needed.

Level A studies do not result in specific recommendations for water resources development, but rather present forecasts that x cubic meters of water supply storage or flood control storage, for example, will be needed in 10 years, y will be needed in 20 years, and so on. Designs and cost estimates are not detailed.

Level B. – Level B studies (regional or river basin studies) are of reconnaissance level and are more limited in area and carried to greater detail than Level A studies. Level B studies seek to resolve complex long-range problems that usually were identified earlier in framework studies. They develop and recommend action plans and programs. Projects at specific sites are identified and tentatively sized, and associated impacts, benefits, and costs (of reconnaissance quality) are determined. The priority of elements of the overall plan is identified for subsequent Level C studies. Level B studies minimize total costs for developing and managing water resources in a basin and ensure orderly development of the resources in order of priority for meeting needs.

Level C. – These project planning or implementation studies begin with a problem or need that may or may not have been considered in prior Level A or Level B studies. Alternative programs are formulated to solve the problem and are evaluated to determine feasibility of solving the problem. Designs, cost estimates, and estimates of impacts and benefits are addressed in detail. A specific course of action is recommended.

Planning Area Boundaries

Geographic areas, river basins, administrative and political regions (countries, states, counties, districts, etc.) can all overlap. Framework plans developed for regions encompassing two or more river basins are usually sub-divided into river basins for more detailed planning. The other extreme is a river basin, such as the Nile, that encompasses a number of geographic and political boundaries.

A situation exists for international rivers with drainage basins occupying several different countries. Tributaries of the Nile have their headwaters in the Sudan, Zaire, Uganda, Rwanda, Burundi, Kenya, Tanzania, and Ethiopia; the Nile Basin occupies parts of nine countries. The Niger Basin also occupies part of nine countries, with the primary river source in Guinea and flowing through Mali, Niger, and Nigeria.

The river basin is acknowledged to be the most appropriate unit area for water resource planning and development because it is a natural, specifically limited area that acts as a unique hydrologic system. For each river basin we can define precipitation, groundwater resources, and basin outflow.

In the case of large complex basins, however, it is not necessary to plan in detail for the entire drainage area is a single unit. Some tributary basins can be analyzed independently in a more efficient manner, but in analyzing sub-basins account must be taken of water resource development and management in upstream sub-basins as well.

The extent of the planning area for any specific study is usually determined by the problems, needs, and resources under consideration.

NATIONAL OBJECTIVES AND GOALS

National objectives for water resource development are usually very general in nature. In the United States the Water Resources Council's (WRC) Economic and Environmental Principles and Guidelines for Water and Related Land Resources Implementation Studies, issued March 10, 1985, identified only one national objective: "To contribute to national economic development consistent with protecting the nation's environment pursuant to national environmental statutes, applicable executive orders, and other planning requirements."

Goals are more specific than objectives and indicate how the objectives are to be met. In the United States, for example, national economic development is to be achieved by increasing the value of the nation's output of goods and services and by implementing water resources development programs that have the maximum excess benefits over costs. Environmental quality is to be achieved by the management, conservation, creation, restoration, or improvement of the quality of certain natural and cultural resources and ecological systems.

National objectives in developing countries often are more specific and frequently include such objectives as:

- Achieve food self-sufficiency through increased irrigated agricultural production.
- Eradicate poverty and increase employment.
- Redistribute income.
- Encourage regional development.
- Enhance social well-being and quality of life.
- Establish new settlements or new industrial centres.
- Modify population distribution.
- Generate hydroelectric power.
- Improve rivers for dependable navigation.
- Provide safe drinking water.

In the Proceedings of the U.N. Interregional Seminar on River Basin and Interbasin Development, Budapest, 1975, the objectives of water resources development in Bangladesh, for example, are given as:

- Confine river flows to stable and fixed beds at all stages of discharge through embankments and river training.
- Control water flows from river to land.

- Ensure drainage of water from the land into the river.
- Provide irrigation by the coordinated use of surface and ground water to the maximum extent.
- Prevent flooding from the sea through coastal embankments and estuary closures.
- Generate hydropower where feasible.
- Improve river channels for navigation and provide regulated navigation routes.

The same U.N. publication gives two broad objectives for the Malaysian five-year social and economic development plans (which include river basin development):

- Eradicate poverty by raising income levels and increasing employment opportunities for all Malaysians, irrespective of race.
- Accelerate the process of restructuring Malaysian society to correct economic imbalance, so as to reduce and eventually to eliminate the identification of race with economic function.

Planning objectives

Planning objectives are derived from problems, needs, and opportunities in the local study area. They are not identified as specific levels of outputs. Alternative plans formulated later will determine whether (and how well) the objectives can be met. Project outputs will vary with the nature and size of each alternative plan and, therefore, are a product of plan formulation.

In establishing planning objectives, planners must recognize that other concerns may be so important as to impose absolute constraints on the planning process. Such constraints may be of a legal, public policy, social, economic, or environmental nature. For example, providing increased food protection to urban areas at the confluence of the Yuba and Feather Rivers in California by increasing levee height was not acceptable to local people or local government bodies because of extensive damage and loss of life that had occurred in the past when levees failed in the area. The only solution acceptable to local interests was to reduce peak flows by reservoir storage. When such constraints exist, they must be included in establishing planning objectives.

Early in the planning process there are likely to be many objectives of a rather general nature. As planning progresses, the objectives should be continuously reexamined so that a limited number of very specific objectives are used to develop alternative plans.

Broad planning objectives for water and related land resource development include:

- National economic development.
- Enhancement of the environment.
- Improved social well-being.
- National defense.
- Public health and safety.
- Preservation or improvement of water quality.
- Fish and wildlife.
- Recreation
- Preservation of cultural resources.
- Preservation of scenic values.

Typical problems, needs, and opportunities addressed in Level C planning studies in developed countries include:

- Municipal and industrial water supply.
- Irrigation water supply.
- Flood-damage reduction.
- Navigation channels.
- Wildlife conservation, preservation, and enhancement.
- Recreation opportunities.
- Low-flow augmentation.
- Improved water quality.
- Soil conservation.
- Industrial cooling water supply.
- Electric power.

More specific objectives typically identified in project studies for local areas in developing countries include:

- Provide safe drinking water for rural settlements.
- Increase irrigable land.
- Prevent flood damage.
- Provide land drainage.
- Provide rural, municipal, and industrial water supplies.
- Provide hydroelectric power for rural settlements and development.

- Inject cash into circulation.
- Provide training.
- Provide jobs.
- Provide impetus for development.

Public Involvement in Identifying Objectives

If benefits projected in planning studies are to be realized, particularly for water programs in developing countries, there must be no conflict of interest between the planners and users. Attention must be given to local social and institutional issues, and programs must be responsive to users' needs or resources will be wasted. Planners must recognize the importance of institutional, educational, and social factors before addressing the purely technical problems and solutions. In many instances in rural areas in developing countries it is necessary to build a local social unit to speak for the people as well as to assume responsibility for operation and maintenance of facilities.

The perception of development programs and projects frequently differs between planners and users. For any development program to be successful, it is crucial that:

1. Planners and users have the same objectives.

2. Planners recognize, understand, and respect the values, habits, and interests of the users.

3. The program involves appropriate technology and users understand that any new technologies will be socially and economically beneficial to them.

4. No rapid modification of the user's litelstyle be required, but rather changes and advanced technology be introduced slowly and on a scale with which the users can cope.

Locai populations must be given the opportunity to participate in identifying problems and needs and must have a voice in design and operation of development programs. Special effort may be required to involve less powerful groups, such as women and children. For example, if irrigation is a project purpose, planners must have a clear understanding of the role of women and children in agriculture because they constitute the majority of most rural populations. Women's real economic and social activities (including farming, supplying domestic water, etc.) must be understood and evaluated by the planners rather than simplistic acceptance of preconceived ideas of their role. Above all, development programs must be compatible with local conditions.

Public views are often the views of small special-interest groups and can be informed, rational, creative, and very helpful in the planning process. They can also be uninformed, emotional, irrational, obstructive, and harmful. As a whole, the public usually has difficulty envisaging what will probably happen over the next 10, 50, or 100 years without a project as well as with a project. The public will seldom have much to say, will tend to believe they do not know enough to comment, and will be unable to make meaningful suggestions unless led by a smaller active group especially concerned for some particular reason. Therefore, it is incumbent on planners at an early stage of the planning process, through a program of public involvement, to obtain an understanding of public concerns and problems and to inform the public about what the situation is and probably will be in the future without a project as well as what a project could do to make things better, (Petersen, 1984).

Appropriate public involvement techniques usually differ from one planning study to the next. In the United States a rather formalized procedure has been developed over the past 20 years to obtain public views that may include:

1. Formal public participation
   - Public hearings
   - Public meetings
   - Brochures

2. Citizen advisory committees

3. Informal coordination
   - Meetings
   - Workshops
   - Interviews

4. Media coverage
   - Newspapers
   - Radio

        – Television

        – Audio and video cassettes

5. Formal coordination with
   - Concerned Federal agencies
   - State and local agencies
   - Special-interest organizations

Public involvement programs usually have four primary general objectives:

1. Identifying local needs, preferences, and priorities.
2. Identifying all impacts of alternative and recommended programs.
3. Promoting understanding and support for the identified objectives and solutions proposed.
4. Providing opportunity for those affected by water resource development to influence decisions regarding development.

Public views are probably both more important and more difficult to obtain in rural areas in developing countries. The difficulties stemming from lack of communication are primarily due to:

1. The dichotomy between national and local priorities.
2. Much of the water resource planning affecting rural areas is conducted by urban personnel of the national government.
3. Much of the water resources planning is conducted by or jointly with foreign agencies and foreign private consultants.

A representative case that illustrates the importance of these factors in success or failure of a water development program is that of the Awash Valley program in Ethiopia. Winid (1981) states that the Ethiopian government had not clearly defined goals for the program and there was ambiguity as to whether the program was primarily to increase agricultural production (with emphasis on crops for export) or should give priority to settlement and sedentarization. Planning for the Awash Valley was done by a foreign consultant, and Winid attributes lack of success for that development to a variety of factors, including:

1. Lack of knowledge on the part of the foreign consultant about political, economic, social, cultural, scientific, and technical conditions in the study area because of:

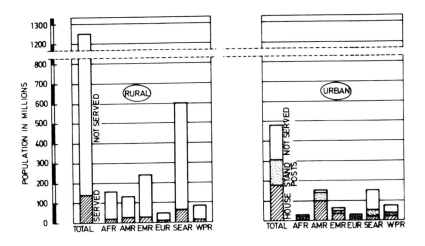

Fig. 1.6 Urban and Rural Population Served with Water in Developing
Countries (WHO Member States), 1970
(Source: van Damma, WHO International Reference Center for
Community Supply, 1973)
By region: AFR – Africa, AMR – Americas, EMR – Eastern
Mediterranean, EUR – Europe, SEAR – Southeast Asia, WPR –
Western Pacific.

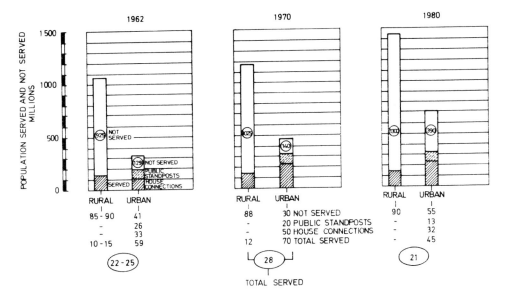

Fig. 1.7 Trends in Community Water–Supply Situation in Developing
Countries (WHO Member States), 1962–1980
(Source: van Damma, WHO International Reference Center for
Community Supply, 1973)

- The short period (four years) they were employed.
- Their limited contact with a wide sector of local people.
- Frequent changes of advisers having different views.
- Ineffective supervision because local counterparts were too young.
- Relations with the UN bureacracy.
- Introduction of extensive new techniques and excessive investment.

2. Lack of awareness by foreign advisers of the true political and governmental obstacles to development.

3. The fact that the Advisers' plans:

- Were not nationally integrated and were divided between specialized UN agencies.
- Were focused on the mining and consumptive industries.
- Were export-orientated in agricultural production.
- Did not relate to existing locally trained manpower.

4. Locational policies for industry were largely drawn up in the national capitol and other large cities.

On the other hand public participation in poor and backward countries can have difficulties due to:

- Lack of understanding.
- Desparation, e.g. starvation.
- Short term objectives taking precedence over long range objectives.
- Corruption causing subjective decision making.

Project Scale and Appropriate Technology in Identifying Objectives

In developing countries there is usually a need for both large multiple-purpose water resources projects to develop potentially fertile regions of the country and for small projects affecting small local areas. Without Hoover Dam, constructed some 50 years ago in the United States, it is highly unlikely that the southwestern part of the U.S. would support the population centres and agricultural and industrial complexes it does today. Many developing countries need such large-scale projects. However, outputs from large, complex projects must be made available to users (such as farmers) through small operational units which they can

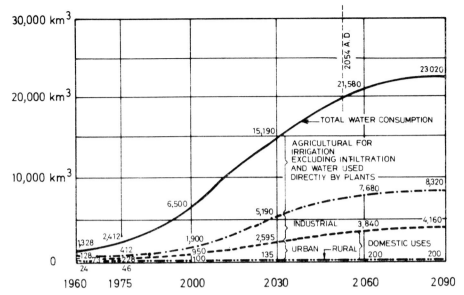

Fig. 1.8   Projection of World-Wide Total Water Demand
(Source: Doxiadis, Water for Peace, 1967)
[in km^3 = 10^{9}m^3 per year]

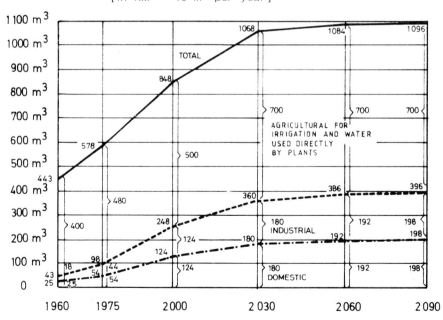

Fig. 1.9   Projection of World-Wide Per Capita Water Demand
(Source: Doxiadis, Water for Peace, 1967)
[in m^3/capita/year]

understand and with which they can interact. Small hydropower developments and provision of safe drinking water can be very effective in improving health and quality of life in rural areas.

For projects that are designed to meet specific needs, size may be determined from the magnitude of needs to be met, for example:

- Volume of reservoir storage for flood damage reduction.
- Volume of reservoir storage or ground-water pumping for irrigation water supply.
- Kilowatt-hours of hydroelectric generation.

For other projects size may be related to limitations of the specific site. For example, maximum storage at a given reservoir site.

Size may also be determined by an administrative or political decision:

- To build a hydroelectric power plant of a particular size.
- To provide irrigation water to bring a specific area into production.
- To provide safe drinking water to villages in a given area.

Finally, in the United States, size of a project is usually determined by analysis of benefits and costs of the "last-added increment." In this analysis increments are added (for example, volume of flood control space in a reservoir), and average annual costs and benefits for each increment are computed. Increments for which benefits exceed cost are justified additions to the project; the increment for which benefits just equal cost is the last increment to be added to the plan based on economic justification criteria.

The abrupt introduction of high-technology, high-cost measures which directly affect large groups of people (for example, sophisticated farm machinery and irrigation technologies) has proven unsuccessful in many countries. It is recognized that technology associated with water resources development should be appropriate for the users and for local conditions as well as being a suitable technical solution. It is now generally agreed, for example, that to increase food production in developing countries, the focus should be on the small farmer and on gradually improving traditional farming methods. Gradual change is more understandable and more acceptable to rural people and can be facilitated by demonstration projects and educational programs to help them comprehend how proposed changes can improve their well-being, health, and quality of life. Thus, the level of technology associated with a water development plan, need not be either the most advanced available or the traditional technology in the country, but should be at a level with which the users can interact with confidence and which will lead to economic and social betterment.

In the United States the agricultural experiment stations and farms and the agricultural extension service activities of land-grant colleges have contributed greatly to improved agricultural production by the American farmer. Change can often be introduced effectively through the children who are usually receptive to trying new things. In the United States the 4-H and Future Farmer clubs for young people and vocational agricultural high school courses have been the outreach to children of farmers. Similar programs, modified for local conditions, might be effective in many developing countries.

THE IMPORTANCE OF POPULATION PATTERNS

Planners must clearly understand the importance that population growth rates play in a country's ability to achieve national goals such as obtaining self-sufficiency in food production. One of the national goals of the Kingdom of Morocco is food self-sufficiency. Historical population data for Morocco from 1960 to 1980 are shown on the figure below together with three projections for future growth made several years ago with growth rates of 4, 3, and 2 percent.

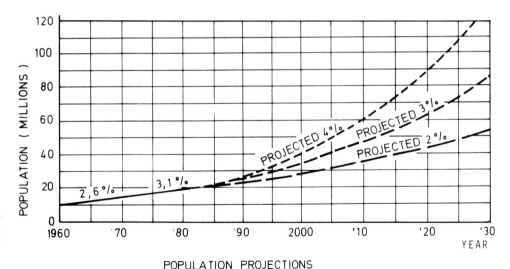

POPULATION PROJECTIONS

Fig. 1.10 Population Projections, Morocco

With a population of about 21 million people in 1982 and a "most likely" future growth rate of 3 percent, Moroccan populations would more than double in 24 years and would increase 8-fold (to 167 million) in 70 years, or within the useful life of any major water development project being planned today.

Kenya has the highest rate of population growth in the world, 4 percent. If that rate continues, their population will double in 18 years, triple in 28 years, increase 5-fold in 40 years, and increase 15-fold in 70 years! These are staggering figures in terms of meeting objectives for major water programs.

The need for several outputs from water development programs (for example, irrigation water supply, municipal water supply, hydroelectric power) is closely related to population. Planners must recognize the relationships between population needs and the importance of such relationships in establishing realistic long-term objectives for water resource development.

FISCAL CONSTRAINTS

The importance of a realistic, fiscally responsible schedule of expenditures cannot be stressed too highly. The costs of almost any nationwide water resource development program will be very high and should be evaluated by comparison with the current division of expenditures in the country's national budget.

In the United States in recent years, approximately 42 percent of the national budget has been payable to individuals under various programs; 29 percent has been for national defence, and only 0.3 of one percent has been for water resources programs. (However, 0.003 of the national budget is approximately $3.6 billion.)

It will often be found that an optimum water resource development program requires an unrealistic share of the total annual national expenditures of a developing country (sometimes as high as 20 percent). Such programs clearly cannot be implemented. The objectives and projects must be scaled back and staged over time so that the development programs are affordable.

TABLE 1.2 World Water Balance, by Continent

(Source: Lvovitch, M.I., EOS, Vol. 54, No. 1, Jan. 1973, Copyright by American Geophysical Union)

Water Balance elements	Europe	Asia	Africa	North America**	South America	Australia²	Total land area*³
Area, millions of km²	9.8	45.0	30.3	20.7	17.8	8.7	132.3
			in mm				
Precipitation	734	726	686	670	648	736	834
Total river runoff	319	293	139	287	583	226	294
Groundwater runoff	109	76	48	84	210	54	90
Surface water runoff	210	217	91	203	373	172	204
Total soil moisture	524	509	595	467	1.275	564	630
Evaporation	415	433	547	383	1.065	510	540
			in km³				
Precipitation	7 165	32 690	20 780	13 910	29 355	6 405	110 303
Total river runoff	3 110	13 190	4 225	5 960	10 380	1 965	38 830
Groundwater runoff	1 065	3 410	1 465	1 740	3 740	465	11 885
Surface water runoff	2 045	9 870	2 760	4 220	6 640	1 500	26 945
Total soil moisture	5 120	22 910	18 020	9 690	22 715	4 905	83 360
Evaporation	4 055	19 500	16 555	7 950	18 975	4 440	71 475
			relative values				
Groundwater runoff as percent of total runoff	34	26	35	32	36	24	31
Coefficient of groundwater discharge into rivers	0.21	0.15	0.08	0.18	0.16	0.10	0.14
Coefficient of runoff	0.43	0.40	0.23	0.31	0.35	0.31	0.36

*   Including Iceland
** Excluding the Canadian achipelago and including Central America
2   Including Tasmania, New Guinea and New Zealand, only within the limits of the continent: P - 440 mm, R - 47 mm.
    U - 7 mm, S - 40 mm, W - 400 mm, E - 393 mm.
3   Excluding Greenland, Canadian archipelago and Antartica

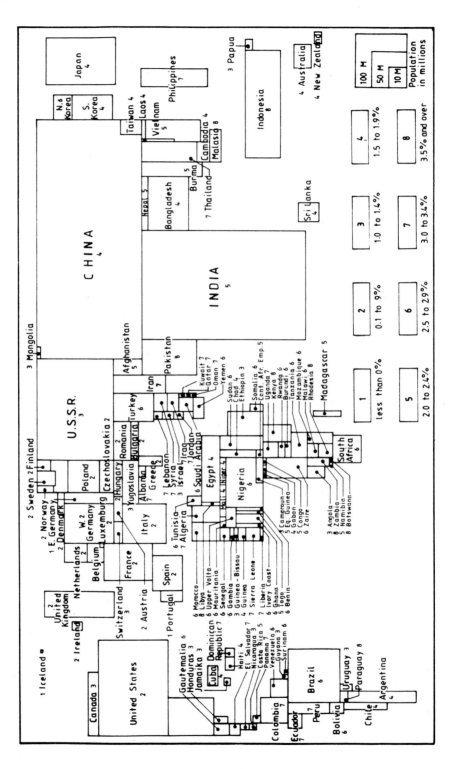

Fig. 1.11   Annual percentage increase in population
(Rates of annual increase based on birth and death rates between 1970 and 1973)

TABLE 1.3 World-Wide Stable Runoff, by Continent

(Source: Lvovitch, M.I., EOS, Vol. 54, No. 1, Jan. 1973. Copyright by American Geophysical Union)

	Stable runoff, km³ [1] p.a.				Total river runoff [2]	Total stable runoff as percent of total river runoff
	Of under-ground origin	Regu-lated by lakes	Regulated by water reservoirs	Total		
Europe	1 065	60	200	1 325	3 110	43
Asia	3 410	35	560	4 005	13 190	30
Africa	1 465	40	400	1 905	4 225	45
North America	1 740	150	490	2 380	5 960	40
South America [3]	3 740	-	160	3 900	10 380	38
Australia	465	-	30	495	1 965	25
Total land area expect polar zones	11 885	285	1 840	14 010	38 830	36

1) Excluding flood flows.
2) Including flood flow.
3) Including Tasmania, New Guinea and New Zealand.

REFERENCES

Lvovitch, M.I., 1973, E.O.S., 54(1) American Geophysical Union.
United Nations, 1975. Proc. Interregional Seminar on River Basin and Interbasin Development, Budapest.
Petersen, M.S., 1984. Water Resource Planning and Development, Prentice Hall.
Van der Leeden, F., 1975. Water Resources of the World, Water Information Centre, N.Y. 1.

CHAPTER 2

SOCIO-ECONOMIC FACTORS

INTRODUCTION

Since the early 1980s it has become increasingly evident to the international assistance and lending agencies that social issues have not been adequately addressed in project planning for water resources develoment and that major unforeseen social and environmental problems have resulted from such programs in less developed countries. The problems have been primarily related to:

1. Resettlement of people living in the project area due to construction.
2. Acceptance of the project and its responsibilities by local people.
3. Inadequate sanitation measures.
4. Water-related diseases.
5. Food production and supply.
6. Ecological change.

The nature and importance of social and environmental impacts in developing areas are different from those in the United States and other industrial countries, and the social and environmental effects are often so closely inter-related as to be inseparable. In some cases such impacts have been so acute that projected benefits have not been realized.

The emphasis of impact studies related to water resources in developing areas has been on economic and, more recently, environmental effects, but social effects have rarely been thoroughly addressed. When social impacts have been studied, the focus has been on populations as a whole, without giving attention to the special role of women as users and conveyors of water. In recent years increasing emphasis has been given to the crucial role of women in planning agricultural development programs because it has become evident that national objectives for food self-sufficiency cannot be met unless the role of women in agricultural production and food processing, preservation, and marketing is taken into account. However, the role of women with relation to water resource development has been largely ignored.

WATER, HEALTH AND ECONOMICS

Water is the key health factor in developing areas. The World Health Organization estimated in 1980 that about 32 percent of the rural population and 73 percent of the urban population in developing countries had access to safe water. About two-thirds of the total population of 4 billion (600 million in urban areas and two billion in rural areas) were without safe drinking and waste disposal. In several countries only a very small percent of the rural people have safe water: 2 percent in Kenya; 3 percent in Gambia, and 5 percent in Zaire, for example.

Water resource development programs have the potential to directly improve the health and socio-economic well-being of people in developing areas. Lack of a reliable and adequate supply of safe drinking water is probably the greatest cause of disease in developing countries, and disease hinders individual productivity and, therefore, economic development.

Traditionally in many rural areas women and girls spend several hours each day fetching the household water supply from natural sources, often from great distances. Access to water in rural areas in many parts of the world is difficult; supplies are frequently polluted; supplies are often limited (sometimes seasonally, sometimes for extended periods of drought); and natural sources are often a considerable distance away.

Typically most developing countries have high rates of population growth accompanied by increasing migration from the countryside to cities as people seek to improve the quality of their lives. The growing urban centres will require greatly increased supplies of safe water in the near future, and in semi-arid areas potential supplies are often severely limited. Also, as an area develops and becomes industrialized, there is increased demand for water, and the associated waste discharge usually leads to increased pollution of water supply sources.

Economic growth and development imply improved living standards for all the people, including better nutrition, better health and health services, better educational opportunities, higher income, and better housing. Few development alternatives have greater potential for improving the health and social well being of people than water supply projects. However, it is often difficult to show project economic justification on the basis of improved health.

The 1980s were designated as the International Drinking Water and Sanitation Decade with the objective of providing safe water and adequate sanitation to all people by the year 2000, but this goal was not met.

In less developed countries, two-thirds of the people still do not have reasonable access to adequate supplies of safe water, and the World Health Organization estimates that 80 percent of all diseases in developing areas is related to unsafe water supplies and inadequate sanitation measures. In such areas water-related diseases contribute to high infant mortality, low life expectancy, and a poor quality of life. Undernutrition and malnutrition clearly reduce the resistance of children to disease and the productivity of adults.

One of the problems in providing safe rural water supplies is that users frequently do not have the capability to operate and maintain the facilities. The technology adopted often is not appropriate for the local culture.

ASSESSING SOCIAL IMPACTS

The following types of impacts related to health and social well-being should be considered when examining potential impacts of alternative water resource programs and recommending programs in developing areas:

1. Impacts on those living in a project area.

   - Changes in communicable disease patterns.
   - Local sanitation problems.
   - Deterioration of water quality (surface and ground water).
   - Adverse impacts on fish and wildlife populations.
   - Lowered nutrition levels due to decrease in per capita food supplies during construction period.
   - Increased employment opportunities with labor-intensive project.

2. Impacts on immigrant project workers.

   - Impaired health due to locally endemic diseases, exposure to toxic chemicals, and physical hazards.
   - Lowered nutrition levels because local food supplies are insufficient for influx of workers.

3. Impacts on those relocated from project area.

   - Problems of ethnicity.
   - Safe water and sanitary measures in relocation areas.

- Compensation for land "in kind." (Land prepared for farming when farmers are resettled.)
- Soil conditions appropriate for same crops as project area.
- Adequate food supplies during resettlement period and until first harvest.
- Access to other towns and health centres.
- Fair compensation for lands.
- Timely relocation.

4. Impacts on health services.

- Greatly increased need for local health services.
- Increased need may be too costly for local resources.

5. Income redistribution.

6. Impacts on living standards.

- Housing
- Availability of safe water.
- Sanitary facilities.
- Electricity.
- Availability of fuel (wood).

ACQUIRING BASIC SOCIAL DATA

Water resource planners must be aware of the advantages associated with involving sociologists and anthropologists to obtain the basic social data needed to evaluate project impacts. Engineers, planners, and economists, unless specially trained, rarely have the background and skills required to obtain adequate and valid social information. Such technical specialists are usually urban males (or expatriates) who have difficulty in communicating with rural people, especially women, and different ethnic groups. International expert teams generally do not include women, and few team members understand or appreciate local customs. Their local informants are usually government officials who, themselves, often do not understand rural people and their culture. Rural people frequently are very cautious in discussions with "outsiders," and the lack of common language and dialect often accentuates communication problems. Planners also must be aware that considerable time may be required to obtain the needed data (especially if the program is large and complex), and the data are often needed early in the planning process.

Economic, social, and cultural characteristics of rural communities and rural people vary widely from country to country and even within a given country. Because of the hierarchical structure of rural people, those in power are usually those who are better off; they are often more open in expressing their views than the rural poor, but their views may differ significantly from those of the majority (especially those of the landless, women, and other groups who are reluctant to oppose the elites.)

PERTINENT SOCIAL EFFECTS

Social factors frequently affected by water resources programs in developing areas can be generally grouped in four categories: (1) socio-economic factors, (2) quality of life indicators, (3) agricultural factors, and (4) services. Potential impacts on these indicators during construction and operation of water development programs must be carefully assessed and considered in planning, design, construction, and operation. The extent to which these potential impacts are positive and the extent to which negative impacts can be mitigated may well determine whether or not outputs and benefits projected for a development program are achieved.

Usually it is much cheaper to incorporate measures to provide a safe water supply, a healthy environment, etc. as a part of project design than it is to add such measures to a project after construction has started.

The following tabulation lists indicators most likely to require evaluation in assessing impacts of a water project in a developing rural area. Conditions vary from country to country and from project to project; thus, the tabulation is not complete (see also Biswas, 1980).

SOCIO-ECONOMIC FACTORS

1. **Quality of life indicators.**

    - Food supplies, food preferences, consumption.
    - Nutritional status.
    - Health, health services.
    - Fertility.
    - Infant and child mortality.
    - Life expectancy.

- Housing.
- Distance to safe drinking water.
- Source and distance to water for laundry and bathing.
- Sanitation facilities.
- Type of fuel, distance to source of supply.
- Electricity.
- Education, literacy, school enrollment.
- Means of transport.

2. **Socio-economic factors.**

- Household composition and demographic characteristics.
- Migration (nomadic, rural to urban).
- Ethnicity.
- Hierarchical village structure.
- Kinship patterns.
- Role of women
- Farm size and type.
- Main economic activities.
- Sensitivity to change.
- Sensitivity to risk.
- Adult employment patterns (male and female).
- Child labour.
- Modification of cultural values and lifestyles.

3. **Agricultural factors.**

- Land fertility.
- Subsistence acreage.
- Cash crop acreage.
- Age of tree crops.
- Animals.
- Fishing.
- Farming tools and equipment.
- Preservation and processing of crops and animal products.
- Role of women and children.
- Government extension services.
- Land inheritance patterns.

4.   Services

- Transportation networks.
- Health services.
- Agricultural extension services.
- Marketing facilities.

ENERGY

For small isolated centres in developing countries, the human and social aspects are the primary considerations for rural electrification. While lighting is the primary need, small agriculturally-related cottage industries are also important. For example, tea processing facilities, saw mills and wood processing facilities, food processing plants, and similar small industries all have the potential to improve economic and social conditions in rural areas.

The World Bank (1980) estimated that households account for an average of about 45 percent of total energy consumption in developing countries, but only 10-20 percent of their commercial energy consumption. In low-income countries, these values are 5 and 10 percent, respectively. Much of the noncommercial energy used by households has limited marketability and is used mainly for cooking. Agricultural production typically accounts for only about 5 percent of a country's commercial energy consumption.

Traditional fuels (firewood, charcoal, crop residues, and animal dung) account for almost all the energy used in rural areas and for about 25 percent of total energy consumption in developing countries. About 75 percent of the population of developing countries (2 billion people) use traditional fuels for cooking. Most of these people have access to firewood, but from 0.5 to 1 billion use agricultural and animal wastes for cooking fuel.

Developing countries have been consuming wood supplies more rapidly than they are renewed. Specific measures to meet the energy needs of rural people are needed as part of any prgram to improve quality of life, including reafforestation and the planting of trees as well as hydropower.

Fuelwood

The energy sources most widely used in rural areas in developing countries are wood, charcoal, crop residue, and animal dung. In poorer countries in Africa these sources supply from 70 to 90 percent of the total energy used. Such sources are especially important in rural areas and among the urban poor even in middle income countries (World Bank, 1980).

Wood has become scarce in many parts of the developing world, and it is estimated that over a billion people have problems in securing adequate fuel supplies. Many villagers who formerly could find fuelwood near their homes now must search for it a half day's walk away. and the urban poor spend a large part of their income on fuel. Thus many developing countries face a secondary energy crises that primarily affects the rural sector of their economy. The magnitude of this fuelwood crisis is immense, and forests of developing countries are being consumed at a rate of 10 to 15 million hectares a year (World Bank, 1980).

The impact of wood scarcity on rural women in developing countries is severe. Rural women and children spend a considerable amount of time (often 4 to 8 hours per day) collecting wood. As supplies become depleted, they must walk farther and farther to gather each day's supply. Water resources planners should recognize the need for fuel supplies and the advantages of including reafforestation and wood lots as planning objectives.

Deforestation is most serious in semi-arid and mountainous areas where it can cause serious problems of erosion, siltation, and desertification. Although the fuelwood crisis is already critical, there are technically and economically sound means for reafforestation. The World Bank estimates that in the order of 50 million hectares of fuelwood would need to be planted in developing areas by the year 2000 to meet the projected need for cooking and heating. The gap between present and required planting levels is particularly large in Africa where it is estimated planting would have to be increased as much as 15 times to meet needs.

Electricity

A very small percent of all village and rural people in developing countries is served by electricity. The World Bank estimated installed generating capacity in developing countries in 1980 at 241 gigawatts, or 12 percent of the world total. Between 1973 and 1978, consumption in those countries grew at an average rate of 8 percent a year, compared with 3.5 percent a year in the industrialized countries. However, their per capita consumption in 1978 was estimated to be only 331 kWh, compared with 6,509 kWh in the developed countries.

Roughly half of the world's hydropower potential is in the developing countries (about 1,200 gW), but only 10 percent has been developed. Many hydro sites that were previously uneconomical have become generally feasible in recent years, but there is a long lead time for large hydro projects. Potential mini-hydro projects that have a shorter lead time are estimated to comprise 5 to 10 percent of the world's total hydro resources; however, their relatively high investment costs may make mini-hydro projects uneconomical for village systems with low load factors. If they can be connected to a power grid, they can be used more effectively.

With the present rate of expansion of rural electrification of about one percent per year, only about 25 percent of all rural people will have electricity by the year 2000. Often electrification of a village does not provide for supplying power to households, but only to pumps, wells, and cottage industries. When power is supplied to households it is often for only 2 or 3 hours in the evening. Construction of small hydropower plants in rural areas is an important means of improving quality of life.

HUMAN DISEASES

Diseases

Access to potable water and availability of water affect public health and basic sanitation furnished to construction workers and resettlement areas as well as other rural people. Principal water-related diseases in developing countries are listed in the following table (Biswas, 1980). At some stage in the life cycle of many disease vectors, water is the breeding or transport medium, as summarized by Petersen, (1984), as follows:

1. Still water and marsh habitat (in lakes, reservoirs, irrigation canals, disposal areas for dredged material, and so on) is the breeding area for mosquitoes that are hosts for malaria, yellow fever, dengue fever, filariasis, and encephalitis.

2. Slow-moving water (in irrigation canals and rivers and along reservoir shorelines) is the breeding area for several species of snails that are hosts for parasites carrying the various forms of shistosomiasis (bilharzia). It is estimated that between 100 and 200 million people in 71 countries are affected by shistomiasis and that 80 percent of the people of Africa are infected. There is an effective drug treatment, hycanthone, but it is very costly. The disease is rarely directly fatal,

but it damages the intestinal tract, lungs, liver, etc. and depresses vitality, contributing to early death.

3. Rapidly flowing water (at natural rapids, steep mountain streams, spillways, stilling basins, and powerplant tailraces) is a breeding area for back flies (Simulium), that are the hosts for onchocerciasis (river blindness). The flies are found up to 15 kilometres from the watercourse. It is estimated that onchocerciasis affects about 50 million people. The major affected area is Africa, but the disease was brought to South America (Columbia and Venezuela) by infected slaves as early as 1590 and to Mexico by Sudanese troops brought in by the French in the 1860s. There are few effective drugs, and they have severe side effects. The parasite can live for 15 to 20 years in humans.

4. Dirty, stagnant, polluted water (in weedy drainage ditches and latrine ditches) is a breeding area for mosquitoes that carry filariasis (elephantiasis).

Table 2.1 Parasitic Diseases

Parasites	Diseases transmitted	Intermediate host	Infection route
Nematoda			
Onchocerca volvulus	River blindness (onchocerciasis	Black fly (Simulium)	Bite
Wuchercira bancrofti	Elephantiasis (filariasis)	Several mosquitoes	Bite
Protozoa			
Plasmodium spp.	Malaria	Anopheles mosquito	Bite
Trypanosoma gambiense	African sleeping sickness	Tsetse fly (Glossina p.)	Bite
Trematoda			
Schistosoma haematobium	Urinary schistosomiasis (bilharziasis)	Aquatic snail (Bulinus)	Percutaneous
Schistosoma mansoni	Intestinal schistosomiasis	Aquatic snails (Biomphloaria; Australorbis)	Percutaneous
Schistosoma aponicum	Visceral schistosomiasis	Amphibious snail (Oncomelania)	Percutaneous
Viruses			
Over 30 mosquito-borne viruses are associated with human infections	Encephalitis; dengue	Several mosquitos	Bite

5. Vegetation near water in humid areas of tropical Africa is a breeding area for the tsetse fly, which is the host for parasites that transmit trypanosomiasis (African sleeping sickness) to people. Although most types of trypanosomiasis are fatal, people are not infected as often as animals. In animals (goats, sheep, camels, mules, pigs, and horses) the parasites cause nagana, a wasting disease that kills the animals; game animals do not appear to suffer from the disease. The problems created by the disease in animals is that the human diet in regions with the tsetse fly lacks sufficient protein (meat, milk), and malnutrition (kwashiorkor, a childhood disease from protein deficiency) is severe. Since the disease prevents the keeping of animals in large tracts of Africa, it restricts the development of mixed farming and prevents human use of large tracts of land, thus contributing to the problem of achieving food self-sufficiency.

Preventative measures

Preventative measures that can be taken to reduce the incidence of the disease vectors and disease discussed above are listed below (Petersen, 1984). If such measures are incorporated in water development programs at the time of formulation and design, control of disease can be much less costly than by later addition of remedial measures.

1. Mosquito population can be minimized by careful design and management measures to avoid creation of mosquito breeding habitat, including:

   Reservoirs

   - Clear reservoir area of all debris prior to filling.
   - Construct drainage channels and grade shoreline (about 1V to 3H) so that all areas drain and no pools are left along the shoreline.
   - Either deepen or fill all shallow areas of the reservoir.
   - Fluctuate the reservoir water surface about 0.3m each week during the mosquito breeding season to strand larvae above the water surface.
   - Gradually draw the reservoir level down during the breeding period.

Irrigation Canals

- Avoid the creation of roadside ditches and other ponding areas.
- Improve irrigation practice to minimize use of water and ponding of excess water.

Many of the above measures for control of mosquito breeding habitat are detrimental to wildlife habitat in general.

The Tennessee Valley Authority was the first agency to develop reservoir operating procedures for malaria prevention. The TVA cyclic pool recession operation is shown schematically in the next figure. The reservoir level is progressively drawn down during the low-flow summer period, and weekly pool level fluctuations destroy mosquito eggs and larvae. Also, small predatory fish and, more recently, Tilapia fish, have been introduced to eat mosquito larvae.

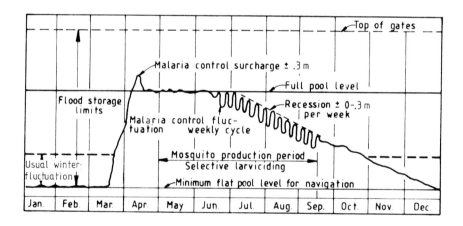

Figure 2.1. Reservoir cycling to eliminate Malaria (Northern hemisphere).

2. Snail populations can be attacked with large-scale applications of molluskacides, such as copper sulphate, but there are adverse effects on fish, microflora, and microfauna. However, fluctuating reservoir levels and rapid drainage of shorelines are effective measures to control them. With rapid drawdown, 50 percent of the snails stranded

will die in a week or two; for slow drying due to evaporation, the period is much longer. A large South American snail that eats the smaller bilharzia snail also has been used to control them. The most effective way to control or minimize snail-related disease is to keep people away from the shoreline by providing clean piped water and efficient sanitation facilities to destroy the human-snail-human chain of infection.

3. The black fly (Simulium) in natural river rapids can be eliminated by "drowning out" (submerging) the rapids or by use of chemicals. On the White Nile 1/40 ppm DDT kept 70 miles of rapids free of the black fly during construction of Owen Dam in Uganda to protect construction forces. Similar treatment was used in construction of Akosombo Dam on the Volta River in Ghana where onchocerciasis was an especially severe problem, and on the Niger River at Kaingi Dam in Nigeria. At Akosombo the reservoir inundated 200 miles of river infested by black flies, but the problem continues in the vicinity of the spillway which discharges continuously. Proper design and operation of spillways and intermittant operation can be effective in controlling black flies in such areas.

4. The tsetse fly (Glossina) breeds in trees near water, in the forest, and in wooded or brush-covered savanna in tropical Africa. It cannot survive in open grassland, and clearing brush is an effective means of control. Some countries have had long-range programs of aerial and ground insecticide spraying and brush clearing. The environmental effects of such programs vary widely, and insecticide use should be properly monitored and controlled. Current research on new techniques include genetic control, traps, and baits.

Aquatic Weeds

Probably the major socio-economic impact of aquatic weeds in reservoirs, watercourses, and canals is that they contribute to the breeding of many disease vectors, including mosquitos, snails, and flies. Little (1969) cites the following problems with aquatic weeds in reservoirs:

1. Blocking navigation channels and making movement of boats difficult.
2. Forming large mats that drift and block power intakes and harbours.
3. Choking tributary streams and irrigation outlets.

4. Forming a dense cover, making fishing difficult or inducing deoxygenation and fish mortality.

5. Reducing the effective capacity of the reservoir.

6. Increasing water losses through evapotranspiration.

7. Reducing recreational utility of the lake.

8. Providing a breeding habitat for disease vectors.

9. Reducing the biologic productivity of the reservoir due to reduced light at surface and subsurface layers.

THE ROLE OF WOMEN

In water resource planning, evaluations have to be made of various social, cultural, economic, and environmental impacts, and tradeoffs have to be made between those impacts that can be evaluated in monetary terms and those that are intangible. To assess impacts requires that planners have an understanding of real-world conditions. Probably the one factor least understood by water resources planners is the real role of women in the use of water and in agriculture in rural societies.

Most water resources development programs in developing areas involve agriculture to a greater or lesser extent, most often irrigated agriculture with the source of supply either stored and managed surface water, ground water, or conjunctive use of both sources. Many developing countries have a common national objective of obtaining food self-sufficiency. If the real role of women in food production, food processing and treatment, food preparation and preservation is not clearly understood by planners, this objective cannot be obtained. Women workers are often undercounted in census data, particularly in the agricultural sector where they are often the predominant labour force. (Women make up as much as 80 percent of the agricultural labour force in some countries.) However, their views are rarely sought in water resources planning.

In developing areas the majority of women are rural women. Typically they are poor, and they are responsible for subsistence agriculture and food production, for fetching water and fuel supplies, and for domestic labour. In some countries they are subject to cultural traditions and restraints, and there are long-standing patterns of women in farming, particularly where the hoe is used. In some areas migration of men to the cities or in search of paid employment on plantations or in mines has left women as head of a high percent of farm households responsible for subsistence agriculture.

The World Bank reports, for example, that in Lesotho over 75 percent of the farming households are headed by women and in Malawi over two-thirds of those working full time in food farming are women. These may however not be representative as many of the males travel to mines in South Africa to work. Households headed by women in developing countries (excluding China) now form, on the average, between 20 and 25 percent of all households except in strongly Islamic societies (Jiggins, 1986).

Recent studies indicate that women are major food producers in terms of volume of food products and hours worked even in countries where they supposedly participate only marginally in farming. Since women produce most of the domestic food supplies, they have an important influence on nutritional levels of the entire rural population and, therefore, on health and productivity as well.

Modernization of agriculture in developing areas has often resulted in increased commercialization of agriculture, with a shift from subsistence farming to cash crops, intensification of production, and technological change.

As production of cash crops increases and mechanization is applied mainly to production of cash crops by men, the demands on women grow to supply more labour for production of cash crops and to produce more of the family food supply as men become less willing or less able to participate in subsistence farming. When machines or other improved technologies are introduced for jobs normally performed by women, men usually take over the new jobs and realize whatever benefits are associated with the new technologies while the jobs women lose may mean their livelihood.

In estimating project benefits, it is relatively easy to determine a monetary value for cash crops, but to estimate monetary benefits for food production by women practicing subsistence agriculture is a far more difficult problem for the engineer or economist.

In many societies, both inheritance of property and the right to accumulate surplus income are difficult for women. Rights of inheritance are usually more restricted for women than for men, and in some areas women have no right to inherit property. Where they have, the right may be nominal, with control of their property transferred to their husbands (or to brothers in matrilineal societies.) Agricultural development programs sometimes contribute to deterioration of women's land rights through requiring legal land title, titles often held only by men.

In developing areas women have been largely ignored in most agricultural extension service programs because the extension staff is typically male and tends to work with larger farms and male farmers. Services and new technologies thus reach women slowly and indirectly, if at all. Women often have little access to various official support services, including not only extension services, but also credit, fertilizers, and new technologies that would enable them to increase production. There are examples of increased general prosperity, but increased malnutrition among women and children when the focus has shifted from subsistence farming to cash crops.

National Policy

Even when national policies dictate a strong commitment to improving quality of life, implementation of such policies is often incomplete. Review of numerous water resource development projects indicates that social factors, including the special role of women, have not been adequately addressed in planning studies, and as a consequence some programs have had unexpected adverse social impacts. Also, some programs could not be successfully implemented because there were no acceptable incentives for the intended users to participate. Planners must use both socio-economic profiles and public participation in the planning of programs to ensure that such development programs are designed to meet the needs and capabilities of the users.

It is important that decision-makers, as well as planners, recognize that many development programs in the past have failed to deal equitably with women; common practice has been to address the needs for the population in general. It is only recently that the international lending agencies have come to understand the high economic cost of not fully utilizing womens' resources. Their contributions are significant. They provide in the order of 50 to 80 percent of agricultural labour; they do most of the subsistence farming; they help produce cash crops; and they participate in harvesting, marketing, and storage of crops. Also, women often control activities that affect family health and, thus, indirectly productivity. There is a very high economic cost to any country that does not use effectively the resources that women represent, and national policies must recognize this fact.

National policy should be to consider the introduction of new technology only following an evaluation of the technology, taking into account national and local needs and existing local social and cultural background. Even well-designed technologies will fail if they are not integrated into the everyday life of the society in which they will be used. If women participate in the management and use of water systems, it is more likely that the systems will be more successful than if they do not. National policies should promote participation by women in planning for development, in assessing alternatives. If some alternatives provide special benefits to women, this should be taken into account; however, if there is a women's component in a project, it should be an integral part of the overall project, not a separate isolated program.

WORLD POPULATION

Concern has been expressed as to the carrying capacity of the earth. Populations continue to grow and the world population in 1980 was estimated to be 5 000 million. The biggest concentrations are in China (1 000 million) and India (700 million) followed by the USSR and USA. However population densities indicate there is little danger of crowding out in our lifetime. That of China (100 per sq km) may be compared with Bangladesh (600 per sq km), which is the highest excluding Singapore and Hong Kong, city countries. There is no clear distinction between developed and developing countries since Holland with a population density of 350 people per sq km is fairly self sufficient and a highly developed country.

The availability of resources for larger populations is only a problem if the countries are to develop to the consumption levels of the first world. The populations of the USA (220 million), Europe (400 million) and Japan (120 million) total only 25% of the world population, so one could expect usage of metals, timber and oil to increase manyfold if the entire world reached the advanced levels of these countries.

The natural check in rate of development is the rate of technology transfer. Even without political barriers there is a generation of teaching required, plus adaptation of ways of life. The cost of such rapid development however could not by borne by the developed countries, and it will have to come largely from within, so that much slower development is possible than we would hope. There is also no ambition, indeed reluctance, of many to 'westernise'.

POVERTY

The extent of poverty in the world is under-appreciated by people of developed communities. Of the world's 5 000 million, less than 25% can be classified as having any wealth. These are predominantly in Europe and North America. An index of wealth can be taken from consumer expenditure and savings patterns. Possession of a family automobile and contribution to some form of pension can be regarded as signs of wealth. On the other hand to be rich implies high expenditure on luxuries e.g. boats, flying, holidays, eating out etc.

Poverty is even more difficult to define. To the western world, Chinese and Soviet citizens appear poor because of their low incomes if translated using common rates of exchange. South America on the other hand has a mixture of wealthy and poor people and the same applies to Africa and Asia where colonization injected pockets of wealth. Within these areas are extremely poor (impoverished) people. Such can be classified as having no hope of earning any money, or even buying anything. They live from hand to mouth, health is poor and clothes, if any, are rags. Probably a quarter of the world's population live like this, largely in Africa and Asia with some groups in South America.

Poverty leads to theft, banditry and illegal activity such as drug production or poaching. Those responsible for supplying arms to many of these countries are to blame for severing progress for many decades. They destroy what was created previously and the countries gradually become unmanageable. This coupled with inept and corrupt governments in the first place is leading to decline of many countries. Perhaps the despair of ever catching up to developed countries, or the taste of wealth by few in power leads to avaricious governments which eliminate democracy to hold on to their declining kingdoms.

Such scenes point to the lack of training provided by previous empires and present developers. Training in methodology, morals and sense of values is needed in parallel. Schooling and working hand in hand are needed for generations before self stimulation the way developed countries want is evident. Unfortunately much development aid is not provided for these reasons. It is often paid to gain access to resources of the developing countries.

With modern communications it is becoming more difficult to brush the reactions and poverty in these countries under the carpet. The consequences will spill over into all countries, affecting quality of life in developed countries. Fortunately reduced international expenditure on defence with the mending of East–West conflicts, may provide the vast resources needed, to set the poorer nations on the path of development. A learning curve will be needed to create such guidance. Many development agencies have experience in training already and their knowledge together within selected investment projects, particularly in the resource field, may speed development.

It has been found that developing communities can learn technical tasks quite easily and take pride in their production and it is hoped more training will be specified in future project specifications. Such training will have to cover a wide field of people, ranging from moral guidance of simple people to technical training of those able to learn and willing to work. Attitude to working days, responsibilities and hard work will also have to be demonstrated, and it may mean sacrifice by the wealthy in accepting these people into their field of concern.

A way of assessing level of civilization could be based on life expectancy which is a function of medical facilities. Fig. 2.3 shows life expectancies around the world.

Human Attitudes

The willingness and ability of the people are necessary to ensure success of a project. Factors which have to be checked in parallel with feasibility studies include:

Government –	policy, integrity, stability
Administrative structure –	systems efficiency, ability of office holders, integrity, experience
Infrastructure –	extent, efficiency
National –	pay levels, standards of living, communication
Workers –	availability, skills, training required, attitude, customs, laziness, pride, superstitions
Backup –	spares, maintenance, operational training
Ouside influence –	theft of produce, equipment, geurilla action, propaganda, aid (can introduce sense of irresponsibility), threats.

Some deficiencies can be tackled within project time frames, e.g. training, but others require decades, e.g. attitudes, or even generations, e.g. customs. If civilization is to be imposed on people which it will in view of increasing population numbers, then it appears training in basic attitudes is the first thing to tackle. The effect of non-engineering factors is illustrated in Fig. 2.2.

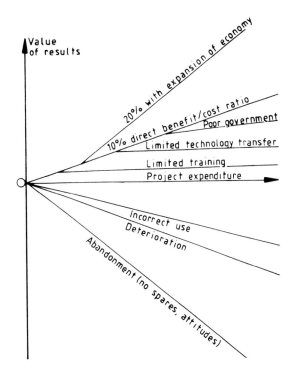

Fig. 2.2   Benefit-Cost relationship with external factors

REFERENCES

Biswas, A.K. 1980. "Environment and Water Development in Third World,"
    ASCE Journal of the Water Resources Planning and Management Division.
Jiggins, Janice, 1986. "Gender-related Impacts and the Work of the
    International Agricultural Research Centre, CIGAR Study Paper No. 17,
    World Bank.
Little, E.C.S., 1969. "Weeds in Man-Made Lakes," in Man-Made Lakes,
    Ghana University Press.
Petersen, M.S., 1984. Water Resources Planning and Development,
    Prentice-Hall.
World Bank, 1980. Energy in Developing Countries.

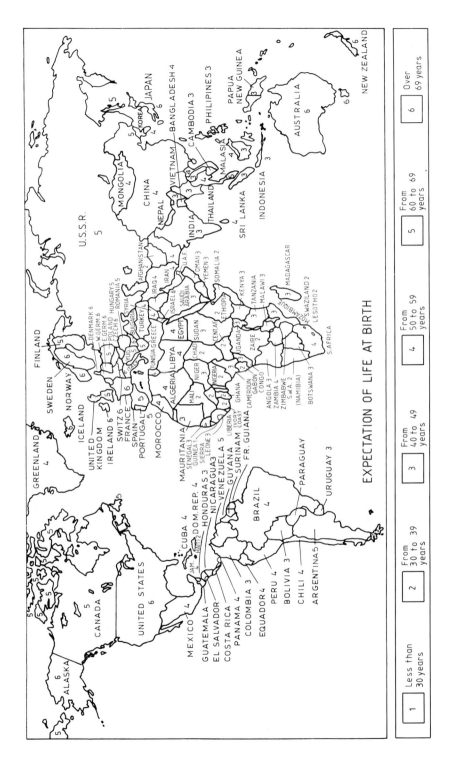

Fig. 2.3 Expectation of Life at Birth

CHAPTER 3

ECONOMIC PRINCIPLES

INTRODUCTION

Water resources development has long been regarded as a way of sparking development and improving sagging or developing economies. The United States itself embarked on major water resources development around the middle of this century in order to provide impetus and stability in the labour market. The World Bank has continued to attach priorities to water resources development in developing countries. Although the emphasis has in recent years swung towards sanitation and water supply this is more from an idealistic and health point of view so do not always provide the same spark that larger water resources development projects do. There have however been many failures in water resource projects, and idle dams and canals provide evidence of ambitious but not pragmatic engineering.

Analysis of the benefits of water resources projects such as irrigation supplies, hydroelectric supply and domestic and industrial water supplies may indicate it is often at first appearance not particularly attractive from the short term economic point of view. In fact using market prices and current interest rates on capital investment, then it would be doubtful whether some irrigation projects could be justified. However, there are many secondary, tertiary and hidden benefits in such development which could be considered in evaluating projects. In addition the economic figures using market prices and current interest rates may be somewhat unrealistic as these figures are distorted by foreign exchange and political attitudes. It is described later how the use of shadow values can ameliorate these effects and result in the use of real values for items such as labour, commodities and even financing.

Later in the book a systematic model for use in planning the development of various levels in a developing country is proposed. The principal used in successive layers of the model is the decomposition of linear programmes developed by Dantzig (1963). Even at that stage successive sub programmes were controlled using shadow values although the modern idea of shadow values is somewhat different.

The top level of the model described herein is that at national level. The second level would be departmental, and the third level in the case of water resources development would be in the nature of river basin

development while the fourth level would be at project level. Each
successively lower model is optimized on its own but using shadow values
on output imposed by the successively higher models. The lower models in
return feed back optimal plans and technical aspects such as output to the
higher model which is further able to refine planning.

This planning process in the theoretical type of model of Dantzig can
be brought to an optimum conclusion with a finite number of iterations of
the successive models. In the case of a real economy both planning and
the economy are in a state of flux and dynamic. A final plan may never
be reached as objectives and funding change with time. The same type of
model as used in the theoretical model can however be used in a
continuously updateable mode for planning national and lower levels of
development as described here.

The model described here is optimized at only two levels, that is basin
development using as an example the Mbashe river basin in Transkei,
Southern Africa, and at the departmental level a comparison of two basin
developments namely the Mbashe basin and selective developments in the
Umzimvubu basin, also in Transkei.

Master Plan Approach

If it is accepted that orderly development must take place within a
national framework then a national policy must be developed. That is,
priorities must be established together with target growth rates which will
be linked to monetary supplies.

Competing departments in the field of development will consider

Environment and conservation

Agriculture and forestry

Water and other resources

Education and training

Urban and rural infrastructure e.g. water, electricity,
communications and security

Trade and industry

Regional development plans e.g. river basins, should proceed within
the national framework. To ensure the correct balance shadow values must
be placed on output e.g. agriculture, water supply, or training. Many
potential basin and project benefits will not appear worthwhile during
project evaluations unless this is done. Thus four levels of planning are
envisaged:

National

Departmental

Regional

Project

Each level will impose shadow values on output of lower levels. The lower levels e.g. project studies will in turn be fed up to the superior level which will compare alternative plans, and possibly revise shadow values.

ECONOMIC FACTORS

Discount Rates

Comparison of capital expenditure and time dependent income can be made in various ways and using alternative discount rates. The different methods all involve comparing incoming cash flows at some time with outgoing cash flows (expenditure) at possibly some other time. The operation of how to compare year 'A' dollars with year 'B' dollars is the one which is most vexing. The following factors affect the comparison:

i)   Interest can be made on early investments so one dollar this year will produce $1 + i$ dollars next year, $(1 + i)^2$ the year after and so on, where i is the annual interest rate as a fraction.

ii)   Inflation lowers the value of one dollar next year so it may only purchase $1/(1 + f)$ worth next year assuming it purchased 1 dollar worth this year, where f is the inflation rate.

iii)   There is a time rate of preference for money. That is, one dollar is more sought after this year than next year. The present preference is influenced by risk of losing the dollar if payment is delayed, lost investment opportunities, the urge to live for today rather than tomorrow, and unknowns in the way of tomorrow's prices, world conditions etc.

iv)   Governments and their treasuries may attempt to regulate economic growth by controlling interest rates. They may adjust interest rates to control inflation, to induce inflow of capital, to provide stable conditions, to provide employment, or to control exchange rates or import-export related cash transactions.

In order to solve the problems of both including inflation and selecting a discount rate, the following has been proposed:

(i)   No inflationary component is included in either prices or interest rates i.e. they are in real terms. This is equivalent to taking a loan and purchasing capital equipment from a country which has no inflation. A real interest rate of 6% p.a. has been proposed by the World Bank for this purpose. This does seem high in Africa and the developing countries where inflation rate can approach or even exceed local interest rates.

or

(ii)   Inflation is included in both price and interest rate projections. Nominal rates are thus used. It may in fact be easier to project inflation in prices than interest rates in developing countries, since inflation is linked to world prices, whereas interest rates can be controlled and varied by governments and they may be artificial, i.e. not reflect the real value of money.

In the case of private investors which operate on a financial basis, they would make the best estimate of interest rates with a marginal or sensitivity analysis for alternative rates. In the case of governments, the interest rate to use will also express the desire to provide immediate or long term opportunities. Thus a low interest rate would favour immediate expenditure and development. A high interest rate would disfavour development, but attempt to curb inflation by creating a shortage of capital.

In many developing countries the interest rates are artificially low i.e. below current inflation rates. This is an attempt to encourage local development but it would only work in closed economies e.g. South Africa, since otherwise capital would flow out to countries offering high real rates of interest. Low interest rates and high inflation rates also discourage savings and waste saved capital, i.e. they are short term solutions resulting in long term problems.

In general developing countries require high capital expenditure to create employment and infrastructure. They can do this at the cost of high inflation rates i.e. create artificial money, provided the economy does advance in order to pay for the early folly of over-expenditure. For instance the U.S. under President Reagan worked on a planned deficit budget, but this broke down when growth slowed down.

Inflation-fueled development and a changing economy can be wasteful and discourage saving and long term growth. It in effect robs Peter to pay Paul. It will also adversely affect foreign exchange rates and prices of imported goods will rise due to devaluation of local currency. It will discourage foreign investment however, so except in scenes which are affected by political events, inflation spending is to be discouraged. In some countries low interest rates will induce high internal expenditure which will assist in reducing outflow of capital. Since little foreign inflow of capital can be expected anyway, the low interest rates are not all that important. They could be expected to rise however when political stability is evident.

An alternative, (preferred from the model point of view) to using artificial interest rates as a way of controlling development, would be to use shadow values on output and other benefits e.g. employment opportunities. This subject is discussed in detail later.

In developing countries development decisions are usually based on comparison of benefits and costs. In particular, water resource agencies prepare feasibility reports summarizing benefits and cost of projects. Projects are either accepted or rejected depending on the ratio or difference of benefits and costs. Benefits are counted in terms of the monetary value of products or services resulting from a project, and costs comprise the investment in construction and development. It is implicit that market prices adequately reflect values of commodities. The sensitivity of net benefits to input data may, however, not always be fully appreciated. The analysis may be vulnerable to factors such as subsidies, discount rates and foreign exchange.

In undeveloped and developing countries this type of analysis i.e. benefit/cost comparison may break down. The following factors could cause the evaluation of benefits in monetary terms to be unrealistic:

i) While in First World countries the price system may be fairly stable, except for the inflation, in developing countries there may be a lag between development and price adjustment. The price structure therefore includes a kinetic or anticipatory component.

ii) Tariffs and subsidies are fairly predominant in developing countries. These may be imposed in order to ensure basic requirements of the sub-economic members of the society.

iii)   In undeveloped regions it may be difficult to draw a line between costs and benefits. For instance should provision of employment opportunity on project construction be regarded as a cost?

iv)   A large proportion of income in developing countries is from agricultural produce, which has a very elastic price structure. Deficits or surpluses cause large price changes.

v)   Interconnectedness between economic sectors, particularly in non-free economies, may make the product of one dependent on investment in others. Consequently prices are affected by outside sectors and may not reflect true values (King, 1967)..

BENEFITS AND COSTS

The evaluation of water development projects has been the subject of much investigation, particularly in the U.S. Benefit – cost analysis using net differences or ratios has been the subject of manuals by various Federal agencies. Guidelines on evaluating benefits e.g. due to recreation, and whether to include net benefits only, or add gross benefits and compare with gross costs make project evaluation fairly standard in the U.S. (James and Lee, 1971).

When it comes to developing countries (Baum, 1985), the evaluation of benefits is more complex. The provision of employment, even on construction, and the evaluation of on-the-job training, are difficult to evaluate numerically. General guides are therefore desireable and again evaluation may be made simple using artificial interest rates or shadow values. The latter would appear preferable as the former would weigh all expenditure the same, e.g. imported capital equipment favoured as much as labour-intensive construction.

Provision of short term employment on construction may have debateable benefits. The cost of training and standard of output make it not worthwhile. Long term employment opportunity would seem a better investment, since training costs would be proportionally less, and this will sharpen the edges of workers who may invest their money in profit-producing enterprises e.g. business. Commerce, agriculture and industry seem to be the future sources of income and growth in countries, but the growth cycle needs impetus in the way of injected money and infrastructure which water resources projects can do.

SHADOW PRICING

The United States Water Resources Council, and the World Bank and the United States Agency for International Development have developed a basis for pricing hidden items which could influence the viability of water resources projects. It has been proposed that shadow prices or accounting prices be applied to various commodities in order to produce the true value to the economy. In our model the shadow prices are imposed on the current program by the successively higher master program. The shadow values are, in fact, generated by the master program and added to the actual prices of commodities or construction.

Shadow values are used in the computer model primarily for adding to the benefits from commodities produced by water resources development such as agricultural output or electricity or even rural water supply. The use of shadow values extends further than this, however, and it can also be used to affect the capital cost of a scheme. For instance under–utilised labour being employed on such a project has a shadow value on the economy and, therefore, that should be subtracted from the cost of the project. On an even wider scale shadow values can be used to correct for incorrect rates of exchange.

Many developing countries have severe trade restrictions and exchange control which distort values. Commodities produced locally are thus favoured on the one hand by import duties but prejudiced when they are exported owing to the artificially high price if the currency is bouyed. In a completely free economy the value of the currency may drop resulting in the commodities being valued less than in the artificial controlled economy.

It is for the above reason that many funding agencies prefer to work in terms of real discount rates when evaluating such projects. That is, the rate of interest or discount rate used in comparisons of capital and annual cash flows are taken somewhat arbitrarily based on the value of capital in countries with low inflation rates. The inflation rate in developing countries is, however, usually higher than developed countries because of shortage of commodities and forced stimulation of the economy. For this reason real interest rates are considerably lower. Because of the complexity in trying to evaluate projects with high interest and inflation rates economists or planning engineers may have a tendency to prefer the use of the so called real discount rate from a point of view of expediency rather than true answers. Alternatively margin analysis is performed to investigate rates at which a project would be viable.

Development in developing countries is also essential for stability. Employment keeps people active and provides objectives. Expenditure and development is therefore a necessity from the social and security point of view.

Shadow Pricing on Project Costs (Sutherland, 1988)

The question of whether employment of otherwise under-employed people on project construction is a cost or a benefit has been considered by various authors. The subtraction of all labour costs in developing countries is obviously unrealistic as it would considerably enhance the viabilities of many schemes which may otherwise be of dubious value. On the other hand it must be recognized that employment opportunity even for a short period of construction has a benefit. It puts money into the economy which is recycled, albeit on a small scale and short term initially, but once in the system can be used to generate further opportunities.

The Agency for International Development (AID, 1971) recommended that only the following cost of labour be included:

In evaluating whether labour input to a project is a cost, the AID recommends the net cost of unskilled labour taken as his present output. Thus if a labourer is not employed his employment cost is nil. Unemployed workers have a zero marginal opportunity cost, and for the underemployed it is the reduction in value of output if he were withdrawn from his present employment. On this basis, assuming the labour cost component of a project were for example 20 percent, of which 50% were previously unemployed, the evaluation cost of the project would be 10% less than its actual cost. In fact a sightly lower figure may be used to remove subjectiveness.

Shadow Prices on Finance

The rate of return for use in discounting projects was discussed previously. It is proposed that not only internal rates of return be used in evaluating water resources projects but also opportunity costs of capital instead of the actual borrowing rate. This amounts to considering a shadow price on capital.

There are two schools of thought using interest rates as described above. The most basic is to use the actual market rate or the closest approximation which can be made to it in the long term, and the other extreme is to use a social time rate of preference plus a shadow price on capital. The former may be more applicable in developed economies where there is completely free trade, whereas the latter would be preferable where the economy is being put into a state of forced stimulation.

The use of shadow prices can go further than purely influencing rate of discount. It can be used on foreign exchange items to influence local commodity useage. The fact that artificial prices are often used with local commodities may be demonstrated if one realized that customs duties and import duties are added and these are intended to favour local materials. However if the economy were completely free, local production would be stifled and local economy would slump. If the correct import duty has been imposed it may have the same effect as applying shadow values to local production. The AID recommend that, where local prices are highly distorted, world prices be used instead of domestic prices. This is based on the assumption that world prices more nearly reflect the true value. In this case however the correct shadow price must be selected for foreign exchange in order not to bias the analysis. Goodman (1984) gives various examples of the use of market values.

SHADOW PRICING OF WATER SUPPLY (Sutherland, 1988)

Shadow pricing is one of the most mis-understood and mis-used of all of the development planning techniques available. This is probably due to the fact that application of shadow prices is not necessary in the highly developed countries, from which most of our research information has emanated up to now. It is therefore of a high priority that engineers, economists and planners, of countries requiring the application of shadow prices, become more proficient in their calculation and application.

The difficulties encountered in determining and using shadow prices in project analysis must be overcome as the distortions which can result due to neglect of shadow prices in developing regions are too serious to ignore.

The advent of 'Artificial Intelligence' and specifically 'Expert Systems', which are affordable and easily programmed, could well result in a standardised method for the establishment of shadow prices.

Definition :

Shadow pricing is a method of accounting for hidden social and market conditions, which could have an effect on the viability of proposed water resources projects.

The factors most prominant in developing regions and which have a marked effect on prices include the following :

- high unemployment
- poor foreign exchange
- low standard of living
- poor infrastructure
- low standard of education
- markets protected by taxes

For a true analysis of a project's economic viability all factors should be evaluated relative to their effect on the prices to be utilized. Shadow prices are therefore obtained which can either be added to the initial cost or replace them, depending on the method of calculation. Ratios which adjust the true cost to account for external influences can also be referred to as shadow values.

Note that the value obtained has no bearing on a financial analysis as it does not represent the actual money which will be paid for the service. It is an entirely economic price representing the economic cost for providing the service. This will be explained further through the use of examples.

UNDERDEVELOPED COUNTRIES :

The distinction between developed and underdeveloped countries has been the subject of debate for many years. It is not within our field to produce any such classification, yet it is important to the topic.

Jalee (1969) divided the world as follows.

Socialist countries - U.S.S.R. and it's sattelite states in Europe, China, Mongolia, North Korea, North Vietnam and Cuba.

Capitalist countries -

1. Developed - U.S.A., Canada, Europe (excluding countries mentioned above), Japan, Israel, Australia and New Zealand.

2. Underdeveloped - the Americas (excluding above), all of Africa, Asia (excluding above) and Oceania (excluding Australia and New Zealand).

Jalee further pointed out that the Third World countries covered 51% of the world's dry land and contained 47% of the total population at that point in time. The population growth rate in Third World countries was also estimated to be twice as high as for the developed countries.

These data are probably outdated. The highly developed countries have decreased their growth rate, whilst certain of the Third World areas have grown alarmingly in only the last decade. However, for our purpose it is quite reasonable to stay with the groupings set out by Jalee. Development in more than half the world, therefore, requires the application of the techniques we are discussing in order to evaluate development projects on a true economic basis. It can be seen that the application of shadow pricing techniques should be widespread and well documented, however, this is not the case.

Calculation :

To illustrate the shadow price concept consider the following simple example. A farmer has various plots of land available, with projected profits and related man-years as shown in Table 3.1

Table 3.1 Co-operative farm data - Land, Labour and Profits.

Plot no.	Profit per year (dollar)	Labour reqd. (man-years)	Profit per man-year (dollar)
1	350	1.9	184
2	600	2.2	272
3	400	1.9	210
4	100	1.0	100
5	500	3.0	167

If the farmer's labour resource is restricted to six labourers, which plots of land should he farm?

This first step is rather simple and we have all dealt with a similar situation. It is obvious that the farmer must use his scarce labour firstly on the most profitable plot of land and then allocate remaining labour to the next most profitable plot until no more labour is available. The result is shown in Table 3.2.

Table 3.2 Relationship between Land, Labour and Profits

Plot no.	Profit per year (dollar)	Labour reqd. (man-years)	Profit per man-year (dollar)
2	272	2.2	2.2
3	210	1.9	4.1
1	184	1.9	6.0

The question now arises, if another plot becomes available what should the profit per man-year be to make it profitable to re-allocate scarce labour from the plots in Table 3.2 to cultivation of the new land. From Table 3.2 we can see that the profit must exceed 184 dollars per man-year to satisfy this requirement. Stating it differently, the farmer, when considering a new plot of land, must compare all the benefits and costs involved, including a shadow price of 184 dollars per man-year for labour to be added to the regular cost of wages.

The shadow price is therefore indicative of the scarcity of a resource. If the farmer in the above example had only 4 labourers then the shadow price of labour for any cempeting project would be 210 dollars per man-year (considerably higher due to the reduced availability of the labour resource).

In the past many economists have advocated the sole use of market prices arguing that market forces will govern. This may be true in a free-economy but in an underdeveloped region the divergence of market values from the true economic prices is too large to ignore.

It is a contention that the shadow price should always be employed when planning projects in under-developed regions of the world. Starting with the market price as a base, attempts should be made to calculate the cost to the economy, of the intangible effects related to the utilization of the resource. We then have the Shadow Cost and can evaluate the shadow price as follows :

Shadow Price = Market Price - Shadow Cost

Problems

The greatest problem related to the calculation of shadow values is the acquisition of data. It is often this lack of readily available data that leads the planner to neglect shadow pricing altogether, in favour of a quicker evaluation. The importance of overcoming this energy barrier in order to collect sufficient data to calculate shadow prices is stressed. Without this extra input the decisions made can be regarded as an inaccurate reflection of the true economic picture.

An example from Goodman (1984, pg. 314) illustrates the difference in values with and without shadow values.

B/C based on financial prices   = 0.74
B/C based on shadow prices   = 1.04

This was an actual economic evaluation of the Varder/Axios project in Yugoslavia-Greece. It is obvious that the application of shadow prices had a dramatic effect on the viability of this project. Although this might not always be the case, it is essential that shadow prices be incorporated in all economic evaluations to avoid an incorrect conclusion.

EXAMPLE :

Background – To calculate the shadow price for supplying of domestic water to the people of an African country, the following information was collected from the Development Bank report (1987):
–   95% of the population can be classified as rural
–   the average household contains approximately 8 persons
–   approximately 41% of the population is economically active
Subsistence farming is therefore of a high priority.

Bembridge (1984), estimated the average per capita income to be $194.88 of which only 10% constituted cash. He further found that cattle, sheep and goats are the largest resource of the population. It is significant that the majority of the household tasks are carried out by woman, so it is their time which will be most affected by the provision of domestic water.

Health – many of the diseases which affect rural populations are due to poor nutrition and sanitation, and unhygenic general living conditions (Bembridge, 1984). It was also found by Stone (1984), that 90% of the households drew water from sources open to contamination by stock. Provision of clean potable water will therefore have a direct effect on the health of the population. Present average use of water is 12 l/person/day, with unlimited supplies of water a maximum of 25 l/person/day would be used. This figure is low compared to world standards (20 l/person/day being considered the minimum) and can be expected to rise rapidly as the standard of living increases.

Estimates of the benefits derived from the supply of clean water are sadly lacking. The World Bank report (1980), put the cost of health care for the average family in a low density country at 5–10% of their income. As stated in the background 90% of diseases are due to contaminated water and we assume that with the supply of 25 l/person/day the health cost per family will be reduced to a related amount. Figure 3.1 shows an approximation of the cost of not supplying clean water.

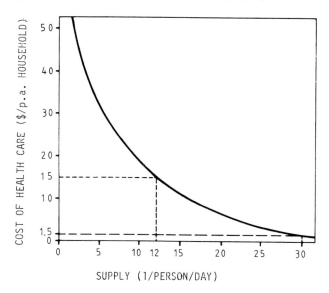

Fig. 3.1 Health costs due to not supplying sufficient water.

Labour – the labour cost will mainly be due to time saved by the elimination of fetching water, which is generally done by women. Bembridge (1984), established that the average household spends 3 hours/day fetching water. Assuming a labour rate of $4/day for a casual labourer (woman) we obtain a cost of:

Cost = 4 × 3/9

= $ 1.33/day/household

This is assumed to apply to the present average usage of 12 l/person/day and the costs are shown in Figure 3.2.

Farming – the time saved due to the supply of water is likely to be used in farming endeavours. We will neglect livestock farming as this is mainly carried out by the male members of the family who are not affected directly by not having to collect water. From tables in Bembridge (1984), we find the average benefit due to farming activities is approximately $48 per household, which constitutes 25% of the per capita income.

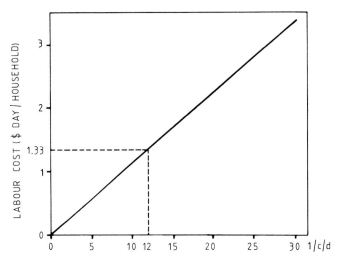

Fig. 3.2 Cost of labour time lost for collection of water by hand

Estimating the time presently associated with farming as 3 hours/day we see that at an average consumption of 12 l/person/day the available time for farming will double. This will not however double the benefit, as the major portion of farming products are for own use. Again from the same tables (Bembridge, 1984), we see that sales of produce constitutes approximately 28% of the benefit. The resulting benefits for various amounts of water supplied are shown in Figure 3.3.

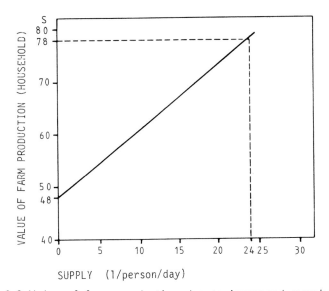

SUPPLY (1/person/day)

Fig 3.3 Value of farm production due to increased supply of water

Summary – the costs evaluated above are summarised in Table 3.3 and plotted in Figure 3.4. This then represents the "Shadow Price" for the supply of domestic water. The final price will be the financial cost less the shadow price evaluated here and will depend on the type of water supply envisaged.

It is obvious that a variety of factors were neglected, but the analysis undertaken did not warrant a deeper investigation. Some of the factors neglected were:

- the benefit of an assured water supply for the farming
- the benefit of the male work force having a higher employment rate based on an estimate of the value of time 'not' sick due to provision of a clean water supply.
- the benefit of reduced government health subsidies due to improved health through the supply of clean water.

TABLE 3.3 Summary of social costs for not supplying domestic water.

Supply (1/pers./day)	0	5	10	15	20	25
Health care cost		8.8	5.2	3.0	1.8	0.7
Water carrying cost	277.0	222.0	166.0	111.0	55.0	0.0
Farm prod. cost	8.6	6.8	5.1	3.4	1.7	0.0
Total social cost of not supplying water		237.6	176.3	117.4	58.5	0.7
Social cost (c/kl)		16.2	6.0	2.7	1.0	0.0

(All costs in c/household/day unless otherwise stated)

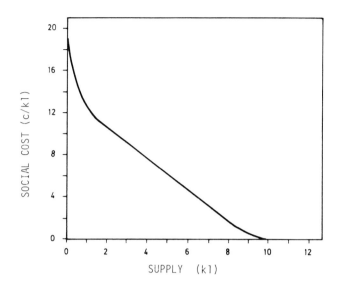

Fig. 3.4 Social cost or Shadow Price for domestic water supply

EXPERT SYSTEM

'Artificial Intelligence' is a new and exciting field of development which holds many possibilities in the field of Water Resources Planning. True 'Artificial Intelligence' is still many years away and even the experimental work is confined to large, main-frame computer systems. What we do have available however, is 'Expert Systems' and 'Intelligent Front-ends'.

The object of both the above systems is to turn the computer into an intelligent machine as opposed to a glorified calculator/typewriter combination. The programmer uses an expert shell, in which he inputs the knowledge of an expert in the particular field desired, in the form of rules. This then is the 'Knowledge Base'. The shell contains an 'Inference Engine' which is able to find it's way through the knowledge base in any direction. The benefits lie in the fact that the expert system can explain to the user 'Why' it requires a specific item of information, or 'How' it has reached a certain conclusion. It is also possible for the system to 'learn' from each interaction with the user and update it's own data files accordingly. In our case a specific region would have it's own expert system for the evaluation of shadow prices, the more it is used the more accurately it can 'infer' the correct information.

Conclusions

Shadow prices are an indispensable tool of economic analysis for regions where free economies do not exist, standard of living is low and economic development is at a very early stage. It is essential that the techniques described in this book be utilised by planners when evaluating new projects in such areas.

Lack of readily available data often leads to abandonment of any attempt to evaluate shadow prices. The example illustrates the complex nature of the required data, but it also shows that the data can be obtained if a thorough investigation into the region in question is made. Acquisition of data in order to carry out calculations of a representative and accurate nature is necessary now, in order to evaluate projects tomorrow.

Expert systems will possibly lead to easier analysis of the factors involved in shadow price determination. This could well lead to a more uniform and general application of the shadow pricing techniques, due to a significant lowering of the energy levels involved and eliminating the need to interpret expert knowledge by individual planners.

INTERNATIONAL FUNDING AGENCIES

Apart from internal budgetary provisions, which in developing countries are relatively small, external financing of the development of the water supply sector in developing countries is now provided from many institutional sources. The principal lending agency to the sector is the World Bank, whilst the regional development banks, namely, the African, Asian, Caribbean, and Inter-American Development Banks now provide significant inputs in terms of loans and technical assistance. Bilateral sources such as the United States of America Agency for International Development (AID), Australian, German, Dutch, Nordic countries, Canadian International Development Agency (CIDA), Overseas Economic Co-operation Fund (OECF) of Japan, and the UK Overseas Development Administration (ODA), are also important contributors in financing projects in the sector.

Recently, new financing agencies, such as the Organization of Petroleum Exporting Countries (OPEC) and the Kuwait Fund for Development, have co-financed projects with the development banks. Several of the United Nations (UN) agencies play an important role in the development of the sector. The principal UN agencies involved in sector development, particularly in water supply are: the World Health

Organization (WHO), which provides technical assistance to developing countries; the United Nations Development Programme (UNDP), which is the co-ordinator of the International Drinking Water Supply and Sanitation Decade (IDWSSD) and which provides grant funds for technical assistance related to the sector in general (e.g. training) and preparation of projects; and the United Nations Children's Fund (UNICEF), which has financed and implemented rural water supply and sanitation schemes in many developing countries.

All these agencies play an active role in the development of the sector, and although there have been cases of overlapping, the various agencies, by and large, have defined distinct and useful roles for themselves. (IWES 1983, Ch. 4,Hollingworth, 1988).

Each of these agencies has a different approach and criteria in the selection of projects for which financing is considered. However, there is a general theme in that each agency does, to a varying degree, subject a project proposed for financing to an appraisal (UN, 1968).

The World Bank

A general statement of the nature and objectives of the World Bank follows plus the annual lending programme as in World Bank Report (1980).

It is a characteristic of all of the above mentioned institutions, that their developmental role is not confined to project lendings. Particular attention should be paid to the diverse nature of the lending instruments.

The World Bank and IFC

The expression, "The World Bank," as used in their Annual Report, means both the International Bank for Reconstruction and Development (IBRD) and its affiliate, the International Development Association (IDA). The IBRD has a second affiliate, the International Finance Corporation (IFC).

The common objective of these institutions is to help raise standards of living in developing countries by channeling financial resources from developed countries to the developing world.

The IBRD, established in 1945, is owned by the governments of 148 countries. The IBRD, whose capital is subscribed by its member countries, finances its lending operations primarily from its own borrowings in the world capital markets. A substantial contribution to the IBRD's resources also comes from its retained earnings and the flow of repayments on its loans. IBRD loans generally have a grace period of five years and are repayable over twenty years or less. They are directed toward developing countries at more advanced stages of economic and social growth. The interest rate the IBRD charges on its loans is calculated in accordance with a guideline related to its cost of borrowing.

The IBRD's charter spells out certain basic rules that govern its operations. It must lend only for productive purposes and must stimulate economic growth in the developing countries where it lends. It must pay due regard to the prospects of repayment. Each loan is made to a government or must be guaranteed by the government concerned. The use of loans cannot be restricted to purchases in any particular member country. The IBRD's decisions to lend must be based on economic considerations.

The International Development Association was established in 1960 to provide assistance for the same purposes as the IBRD, but primarily in the poorer developing countries and on terms that would bear less heavily on their balance of payments than IBRD loans. IDA's assistance is, therefore, concentrated on the very poor countries – those with an annual per capita gross national product of less than $791 (in 1983 dollars). More than fifty countries are eligible under this criterion.

Membership in IDA is open to all members of the IBRD, and 133 of them have joined to date. The funds used by IDA, called credits to distinguish them from IBRD loans, come mostly in the form of subscriptions, general replenishements from IDA's more industrialized and developed members, and tranfers from the net earnings of the IBRD. The terms of IDA credits, which are made to governments only, are ten-year grace periods, fifty-year maturities, and no interest.

The International Finance Corporation was established in 1956. Its function is to assist the economic development of less-developed countries by promoting growth in the private sector of their economies and helping to mobilize domestic and foreign capital for this purpose. Membership in the IBRD is a prerequisite for membership in the IFC, which totals 127 countries. Legally and financially, the IFC and the IBRD are separate entities. The IFC has its own operating and legal staff, but draws upon the Bank for administrative and other services.

While the World Bank has traditionally financed all kinds of capital infrastructure such as roads and railways, telecommunications, and ports and power facilities, its development strategy also places an emphasis on investments that can directly affect the well-being of the masses of poor people of developing countries by making them more productive and by integrating them as active partners in the development process.

In recent years, the Bank has stepped up its lending for energy development. Lending for power forms the largest part of the Bank's energy program, but commitments for oil and gas development have shown the greatest increase.

In 1980, the Bank in response to the deteriorating prospects for the developing countries during the 1980s inaugurated a program of structural-adjustment lending. This lending supports programs of specific policy changes and intitutional reforms in developing countries designed to achieve a more efficient use of resources and thereby: (a) contribute to a more sustainable balance of payments in the medium and long term and to the maintenance of growth in the face of severe constraints; and (b) lay the basis for regaining momentum for future growth.

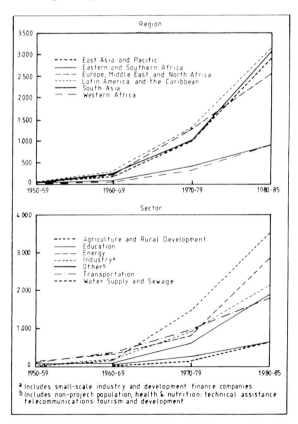

Fig. 3.5   World Bank Average Annual Lending, by Region and
            Sector, Fiscal year 1950–85 (Current US$ millions)

EXPERIENCE IN INVESTMENT BY WORLD BANK (Hollingworth, 1988)

This section focuses on experience of the World Bank in investing in development projects over the past 40 years. Economic, political, and social conditions throughout the world have changed dramatically in that period, and much has been learned about the process of economic development, its complexity, and the pitfalls involved. We understand better today than we did in the 1950s that development is a long, slow process, and one that is often painful. Lessons learned from projects the Bank has financed can be invaluable in planning for future water resource development projects, no matter how they are funded.

Water resources planning has become a very broad discipline in the past twenty years, but it is only recently that we have recognized the importance of considering from the earliest stage of water resource planning the strong inter-relationship between:

- Water resource development.
- Agricultural development (particularly the shift from subsistence farming to cash crops).
- Infrastructure development.
- Modification of environmental and social systems.

In the past planning studies addressed technical and financial factors in detail, with little attention given to cultural variables. It has become obvious that cultural and social factors must be given particular attention from initiation of planning studies.

In 1973 it was estimated that almost 40 percent of those living in developing countries lacked basic human necessities and suffered from malnutrition, disease, and illiteracy. Since then the focus of the World Bank's programs has shifted from comprehensive economic planning toward assisting the poor to increase productivity and increase their access to safe water, health care, and education. These social components have become an integral part of many projects. Since most of the poor in developing contries are rural, the Bank's priorities have shifted to more agricultural projects.

Governments frequently provide irrigation water and electricity for agriculture at little or no cost, at fees that do not cover the cost of even operation and maintenance. Traditionally, governments also provide infrastructure, agricultural extension, health and educational facilities. A large portion of funding for agriculture often goes for irrigation development, but once farming becomes commercialized, infrastructure (such as roads and electricity) is needed. Rural road construction is usually the first requirement, since access to markets is needed for commercial agriculture. Improved roads also encourage those involved in health and education programs to live in villages and provide access to markets. Low agricultural productivity in many developing areas reflects limited investment in rural roads, water, electricity, and so on, as well as agricultural technology. In general, investments in infrastructure in developing countries have been concentrated in urban areas where large numbers of people can be served at low unit cost. Sustained agricultural development requires balanced investments in infrastructure.

Next, a complex network of institutions must be established to manage agricultural development. Establishing skilled staff to manage institutions for planning, decision making, and implementation is difficult and requires much time. The principal institutions needed for agriculture are those concerned with production and distribution; agriculture extension; research; rural finance; storage, marketing, and processing of agricultural products; and transportation. Also needed are institutions concerned with legal land rights, trade, pricing policies, employment policies, and farm cooperatives or associations.

Most governments in developing countries would agree with the five objectives of agricultural development identified by the World Bank (Baum and Tolbert, 1985): growth, sustainability, stability, equity, and efficiency. These are integral to the broad national goal of "food self-sufficiency" of many countries.

- Growth in agriculture is often the primary objective because more food is needed for a growing population.
- Sustainability is maintaining adequate levels of production into the future.
- Stability is important because farm policy must even out the inherent variation of agricultural production due to fluctuating weather patterns and trade cycles.
- Equity involves fair distribution of agricultural benefits among all those involved.
- Efficiency is of particular concern to developing areas because any waste of resources is a real loss to the overall economy, and in agricultural societies such losses can be very large.

The World Bank 1985 Project Review

The World Bank periodically reviews the performance of all projects it has helped finance. The Bank's recent published review was its twelfth annual review conducted in 1985 and released in 1987.

Most of the projects reviewed in 1985 were approved in the second half of the 1970s and completed in the early 1980s. Thus, they were designed in a period of world economic expansion, but completed in a period of economic deterioration when overall growth rates declined, international trade deteriorated, and interest rates rose. This is illustrated by changes in projected and actual commodity prices in the 1976–1985 period shown in the Figure 3.6. The 1985 projects reflect the shift in Bank priorities, with increased emphasis on alleviation of poverty, intensification of small–holder production, and more emphasis on less–privileged groups.

Of the 189 completed projects reviewed in 1985, 80 percent were judged to have been worthwhile, however, in Africa nearly 40 percent of the projects were judged to have unsatisfactory or uncertain results. Both drought and political instability contributed to the low success rate in Africa. The reasons for success or failure are discussed in the following section.

The projects covered a broad spectrum, including 15 power projects and seven water supply/waste disposal projects. While economic return from these projects was generally satisfactory, there were problems with lack of competent and continuous management and qualified staff. There was also a question as to whether the initial benefit stream could be sustained because of excessive transmission losses of energy and water due largely to inadequate maintenance and operation.

Bank reviewers concluded that while the social impacts of water supply/waste disposal projects were difficult to measure, it appeared that their degree of success in reaching particular target groups depended primarily on how carefully target groups had been analyzed initially and on the priority given to reaching them in implementation. The projects were considered to be generally beneficial to the environment, although it was noted that sewerage projects did not keep up with needs due to increased water supplies.

Factors determining Project Success

Many factors determine the success or failure of a project. The Bank's 1985 review indicated that those factors include the following:

1. Project formulation was considered to be critically important, especially the clarity and acceptance of objectives, the technical, administrative, and financial feasibility of the project, and the thoroughness with which the project was developed and appraised. Over one-third of the 189 projects were judged to have been adversely affected by deficiencies in project design or appraisal. Some were based on overly optimistic production targets; for others, the implementation problems faced by the borrower, local institutions, and project beneficiaries were under-estimated.

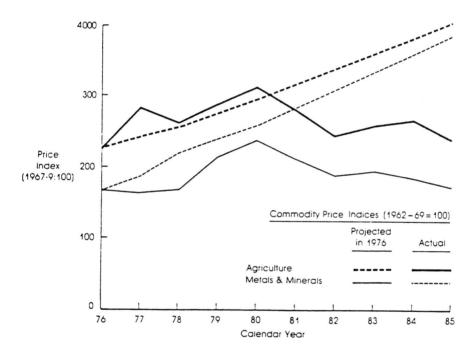

Fig. 3.6 Commodity Prices: Projected and Actual, 1976-85 (World Bank, 1987)

2. Institutional capacity of borrowers is vital in determining the success of projects and the extent to which benefits can be sustained in the future. In the 1985 review, this was identified as the key factor more often than any other. While many of the 1985 projects were effective in strengthening institutions, there did not appear to have been complete understanding of the institutional constraints facing the borrower and of the complex and difficult administrative, political, and cultural problems involved.

3. Strong borrower support is required. Many projects were delayed or scaled down because of a shortage of local funds; others were adversely affected by constraints beyond the control of the project agency, such as manpower shortages and high staff turnover. The borrowing country must take responsibility for its policies and actions and, therefore, must be fully involved in formulation and implementation of a project, but the 1985 review indicated the Bank can sometimes help the borrower to review options and build support and cooperation.

4. External events adversely affected many projects reviewed in 1985. The world recession after 1979 depressed world markets, and in some cases prices were much lower than projected in planning studies. Also a large number of projects were severely affected by the prolonged African drought, political unrest, wars, and administrative changes.

In the 1985 review, the Bank also concluded that some issues identified in prior years continued to be important factors in project success and need continued attention, including:

1. Encouraging environmental impact assessment by project-related dialogue and support.

2. Data on social impact indicators are incomplete, and it is not clear if such information can be gathered economically.

3. The quality and continuity of senior management is vital to project success. While clearer policy and strategy direction are needed, day-to-day intervention in autonomous entities is counter-productive.

4. Training to develop local skilled and managerial personnel is a significant factor in project success. A large proportion of training programs have not succeeded because they were not carefully designed originally.

5. Funding for management, operation, and maintenance is as important as funding for project construction.

About one-third of the projects reviewed were judged to be successes or complete failures. Most projects were typically more successful in meeting some goals than others. Some projects were scaled down during implementation because of funding constraints or limited institutional or administrative capacity.

Economic Impact.

Projects were categorized in three groups in assessing economic benefits:

- Projects for which the economic rate of return (ERR) captures the economic benefit of the investment, principally agricultural, industrial, and transport projects. Comparison of the ERRs estimated in project planning with the 1985 evaluation shows that prior to 1978 re-evaluation values tended to exceed original planning estimates, but that since 1978 the re-evaluation values are less than original estimates, as shown on the Figure 3.7.
- Projects in the utility sector where revenues are used as proxy for economic benefits in estimating ERRs.
- Projects for which no quantifiable indicator of economic return was available at evaluation were assessed subjectively as to likely economic outcome as well as for their achievement of other objectives.

Policy Impact.

About one-third of the projects reviewed in 1985 aimed to support significant policy changes, many of which were related to such areas as agricultural price reforms; energy pricing, conservation, and development; and funding for maintenance. Policy reforms tended to be more complex and time-consuming than expected, and reform objectives sought by the Bank were generally ineffective unless the borrower accepted the need for reform and participated in the process of policy review and formulation. In some cases the policy-making capacity of the country needed strengthening, and this process of institutional development is complex and lengthy.

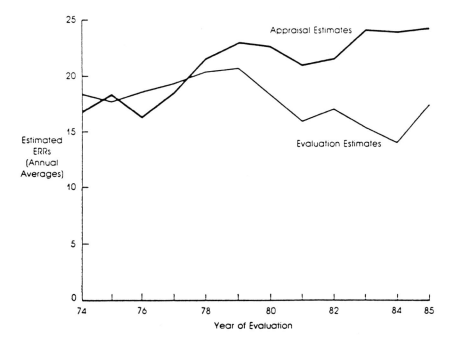

Fig. 3.7 Appraisal and Evaluation Estimates of ERRs, 1974-85 (World Bank, 1987).

Social Impact.

The Bank has been giving increasing attention to social impacts of projects, focusing on reducing socio-economic imbalances, improving quality of life, and enhancing skills and income-earning capability. Most projects with specific social objectives concentrated on education, agriculture, urban water supply, population, health, and nutrition. Project objectives have been designed to enhance living standards by:

- Improving access to and quality of such services as water supply and waste disposal, transportation infrastructure, health facilities, and so on.
- Raising productivity and income through improved education, credit, agricultural inputs, small-scale industrial employment.

The Bank found achievement of social objectives difficult to isolate and measure, in part due to intrinsic difficulties in setting up social evaluation criteria and inadequate data for assessment.

While nearly 17 million rural people benefitted from irrigation, credit, and area development components in the 1985 agricultural projects, achievement of specific social objectives appeared to depend on how carefully the target groups were identified initially and on the priority given to reaching them in project implementation.

Educational components appended to projects without careful preparation and coordination were not fully successful.

Some projects involved redistribution of costs, providing improved services to the poor by cost subsidies by higher income users (i.e. water supply projects in Liberia and Jordan, and power projects in Ghana and Brazil.)

Conclusions on social impacts from the review included:

1.   Assessing the social impact of projects raises difficult questions of methodology and data collection.

2.   Social impacts, especially those involving basic cultural change, take time; such objectives are probably best addressed through a series of projects.

3.   Equity and efficiency considerations sometimes suggest alternative approaches; in such cases both the social and economic costs and benefits must be weighted carefully.

Technological Impact.

Technology transfer has been an important objective for many projects. Of the 1985 agricultural projects reviewed, 44 out of 55 sought to use improved technology to increase production and yield. The technology was generally not new, but was adopted from experience in other areas and was new to the project area.

Experience in agricultural projects showed that to ensure that new technology is adopted, it should have been pretested in the project area and found to be culturally and sociologically acceptable. The Bank found a marked difference between the favourable experience with irrigation projects, where innovation can be readily demonstrated, and the generally unsuccessful experience with other agricultural development projects, especially in Africa, and concluded that more careful preparation could have reduced risks to farmers associated with new technology.

The introduction of sophisticated technology takes time, and it is important to allow time for the learning process. In some cases local agencies are slow to recognize the complexity of the process. There also may be difficulties with spare parts. Even where projects included financing for parts, problems were encountered. Also, in some education programs, special workshops were underutilized because of inadequate stores or parts. There is always the risk that projects may not be adequately maintained or be abandoned after a short time.

The problem of skilled personnel to sustain the new technology is similar. Experience has shown the importance of thorough training to ensure that the technology will continue to be used.

Conclusions regarding technology change were:

1. To be successfully adopted, the technology must be proven to be both sound and appropriate to the needs of the user.

2. It is important to ensure that the borrower can absorb and sustain the improved technology.

3. Due attention must be given to cultural and social factors that might inhibit acceptance of new technology, especially in agriculture.

Environmental Concerns

The 1985 projects faced a number of environmental concerns typical for power and water supply projects. There were two successful resettlement programs in Ghana and Thailand as part of construction of hydroelectric projects. In Zambia, changes in natural stream flow due to construction of a dam and reservoir required augmentation of low flows by special releases from the reservoir. In Thailand, improved downstream conditions due to construction of a hydro project resulted in investment in a controlled water supply system for irrigation.

The Bombay water supply and sewerage project had an environmental impact typical of many similar large-scale projects of this type. Priority was given to alleviating a critical water shortage, and improvement of the wastewater disposal system became increasingly more costly and was considered to be of a lower priority. While it had been recognized that sewerage components of the project could not keep pace with increased wastewater production and that the already extreme pollution in receiving water would worsen, what happened was worse than expected because some of the sewerage components had to be deleted during implementation due to funding limitations.

Sustainability of Project Benefits

The 1985 review indicated that sustaining the estimated benefits of power and water projects appears to require as much attention as implementation of a project. A number of projects had shortages of spare parts (even lack of fuel) because of foreign exchange shortages. This led to maintenance deficiencies and loss of production and distribution capacity. It is at least as important to provide for resources (including foreign exchange) for management, operation, and maintenance as it is for constructing a project.

One of the major concerns regarding sustainability is excessive losses of energy and water in transmission and distribution systems by theft through illegal connections, leakage, or faulty metering. Objectives of 25 percent of the power projects and 45 percent of the water projects included reducing losses, but in most cases results were not satisfactory.

Lessons Learned

Conclusions from the 1985 evaluation for power and water supply/waste disposal projects include:

1. Training programs should be reviewed by specialists to ensure they will be cost effective. Components should be carefully appraised and monitored.

2. Where energy or water losses are problems, projects should include specific programs to deal with the cause. Such programs should be monitored and given as much priority as increasing production.

3. In economies where growth is precarious, emphasis should be on ensuring full use of existing capacity, rather than extension of the systems.

4. With respect to rural projects, it is important that subproject priorities have an economic basis, a proven market with a developed cash economy exists, project designs follow least-cost practice appropriate for the area and the implementing agency has capability to construct and operate the system.

5. More explicit guidelines are needed for evaluating social benefits of such projects to establish a better analytical base for formulating future projects to benefit the rural and urban poor.

6. The Bank can play a needed role in optimizing design standards; using more rigorous economic analysis to examine projects; and coordinating aid from other sources or obtaining co-financing.

7. More should be done to determine the actual coverage of distribution projects and the impact on beneficiaries.

8. An attempt should be made in projects serving urban and rural poor to ascertain that beneficiaries are being reached and to determine how many are affected.

It appears there was little monitoring of environmental impacts of the 1985 power and water projects. A thorough study focusing on the environmental impacts of projects is needed.

Experience with some of the 1985 projects was reflected in follow-on projects, for example:

1. The Mogadishu, Somalia, water supply project was an interim measure while a second project was prepared. Limited capacity of the borrower to absorb organizational and technical improvements and training, identified in the first project, led to adoption of more realistic approaches and timetables for the second project.

2. The Monrovia, Liberia, water supply project generally met its physical objectives, but failed to meet most institutional objectives, and project benefits were rapidly undermined. A new technical assistance project, intended to assist with rehabilitation and improvement of the institution, is realistically based on experience with the first project.

Experience with some projects, other than those evaluated in 1985, illustrates typical problems associated with inadequate regard for existing social and cultural patterns, for example:

1. A south Asian irrigation project to promote cultivation of onions and chillies expected those crops to be fitted into an existing labour-intensive rice-growing system with peak labour requirements at transplanting and harvest times. Those labor peaks competed with the time required for chilli and onion production, and the farmers gave priority to their subsistence crop of rice. Because the cash crops were new to the local culture and conflicted with existing crop priorities and interests of the farmers, they were not adopted.

2. Another Asian irrigation project ignored known social obstacles to forming water users' organizations and relied on the force of ministerial decrees that the farmers refused to follow.

The World Bank completed six detailed evaluations of agricultural programs based on farm surveys as part of the 1985 revue. Surveys of several of those projects that are closely tied to water resource development are summarized on the following pages, (World Bank, 1987) (see also Table 3.4-5).

SOCIAL IMPACT OF MALAYSIA FIRST, SECOND AND THIRD JENGKA TRIANGLE PROJECTS

The three Jengka Triangle projects were the first of six Bank loans to the Government of Malaysia for the development of new lands to be planted to oil palm and rubber and settled by landless people. The projects were part of a large government development and settlement program started in 1956 with the Federal Land Development Authority (FELDA) as executing agency, and which, by the end of 1984, had achieved the settlement of about 89,000 settler families on more than 600,000 ha.

The three projects, approved in 1968, 1970 and 1975, consisted of clearing about 40,000 ha of jungle, planting 26,000 ha of oil palm and 13,800 ha of rubber, construction or expansion of 4 palm oil mills, construction of roads, villages and related social infrastructure and settlement of about 9,200 smallholder families on 4-ha plots.

By the time of the impact evaluation, seventeen years after the first and four years after completion of the third project, two-thirds of the oil palm and about 20% of rubber plantings had reached full production, thus allowing a more accurate estimate of the agricultural, economic, social, financial and institutional impact of the projects.

A sample survey of 229 settlers was carried out in 1985 and follow-up interviews with twenty women settlers were conducted. The survey showed that settlers' incomes are about 3- to 3.5-fold above the rural poverty level and relatively higher for oil palm settlers than for rubber settlers. A large number of commercial activities have been developed by settlers and encouraged by FELDA. However, incomes derived from these activities were lower than expected. The survey showed that settlers' increased incomes had been translated into significant improvement in living conditions. Social infrastructure, particularly education, has been important in both attracting and retaining settlers.

The projects' negative impact on the environment was found to be less severe than expected at project completion. Soil erosion due to land clearing was minimal; all oil palm mills were equipped with efficient treatment plants; there was no indication that climatic change has resulted from the development of Jengka. The clearing of forest land, however, had a considerable effect in terms of reduction of wildlife populations, as protection measures now used were not known at the time the projects were implemented.

Although women play a major role in the agricultural activities, rubber in particular, their contribution is still constrained by traditionalist sentiment.

The financial impact of the projects has been positive for all parties involved in Jengka. While settlers have increased their living standards, FELDA enjoys a healthy financial position and the State of Pahang and the Federal Government have been able to obtain substantial revenues from the projects through land taxes and export duties. The cost recovery rate from settlers has been excellent for oil palm but less satisfactory for rubber, resulting in the need to extend repayment periods beyond the initially planned period for rubber settlers.

SOCIAL IMPACT OF THREE IRRIGATION PROJECTS IN KOREA, TURKEY AND SRI LANKA

In the Korea Pyongtaek-Kumgang Irrigation Project, it was found that project farms were about 17.8% larger than the average national holdings, but farmers only had average incomes 6% above the national level. This disappointing result can be explained by Korea's rapid industrialization and the proximity of Seoul to the project area, which caused a drain on farm manpower and large increases in farm labour wages. Though spreading, farm mechanization has not kept pace with the reduction in farm labour supply. Because of migration to urban areas, fewer young people are now employed in agriculture, while older males increasingly do the work with the help of farm machines and female labourers. Revision of farm size ceilings (now 3 ha) will become essential to prevent further migration. A farm large enough to allow a potential farmer both to have a suitable income and be a part of Korea's modern culture may be necessary to attract younger people back to agriculture.

The Turkey Seyhan Irrigation Project has contributed to significant increases in farm incomes. Living standards greatly improved, demonstrated by better health, education and to a certain extent, lower birth rates. Although land tenure in the project area is highly skewed (23% of the families hold 80% of the land), all smallholders have benefitted from the project. It also helped improve permanent farm labour incomes and provided about 30,000 man-months of employment for seasonal labour.

People in the area are convinced that the project is the source of their great fortune, characterized by a levelling up: poor before, they now are free from debt and risk, they have access to innovation and technology, high living standards, and a stake in the system that only the richest farmers had before.

However, problems loom. There is room for only some children to take over the land, and the fact that good technical education can be rewarded by high farm incomes provides incentives for all children to assert their claim to farm shares. Some technical problems remain unsolved. Complaints about insufficient credit are common. Policies greatly favour farmers, but subsidy policies could change, as the Bank has sought over the years.

The Sri Lanka Lift Irrigation Project failed to provide irrigation water adequately, dependably or equitably. Systems were underdesigned; water supply was inadequate or irregular; canal deliveries were unsynchronized with requirements of lift sytems; broken down pumps took too long to repair; and water distribution was poor. These problems were due to design and implementation deficiencies, some of which were overcome in time, but not until many farmers had justifiably lost confidence in the schemes.

But the technology of chilli and onion cultivation has begun to spread within the project areas and limited export markets for green chillies have developed. Thus, although the market prospects are narrowing, lift irrigation continues to be used by farmers for the cultivation of chillies and other high-value crops. In such cases, irrigation water is likely to supplement rainfall or gravity irrigation and lift mechanisms may continue to be individually controlled.

Project farmers enjoy incomes distinctly higher than non-participating smallholders. However, with no funds to replace pumps/engines in the next two years when their economic life comes to an end, the project may become unsustainable unless farmers themselves, or government, take action to replace the pumps/engines in time. The lined channels, pumphouses, pipes, etc., which have another 15 years of useful life, may become redundant without operating pumps.

THE EXPERIENCE OF THE WORLD BANK
WITH GOVERNMENT-SPONSORED LAND SETTLEMENT

A recent OED study on government-sponsored land settlement is based on 34 completed Bank-assisted projects which had been approved during the period 1961-78. The total cost of these projects was US$1.59 billion, exceeding appraisal estimates by 93%. Total Bank Group lending amounted to US$413 million. Implementation took on average 85 months, or 36% longer than originally expected. The average re-estimated ERR at completion of the projects which had been audited (27 projects) was 15%, compared with 17% estimated at appraisal; 62% had ERRs of 10% or better, and 50% of the successful projects had major multiplier effects.

The findings of the study confirm that successful settlement projects can not only increase agricultural production and benefit large numbers of low-income families, but also catalyze a process of regional development. Such projects have the potential to combine development with sound environmental management. Provided adequate attention is paid to the nature of the settlement process and to a number of key variables, investments in land settlement would appear attractive. Among the key variables deserving special attention are project management, site selection, research and extension, marketing services, phasing of investments, provisions for operation and maintenance, mobilization of settler initiative, and promotion of settler-run organizations.

Concerning Bank performance, it was found that lessons learnt need to be more systematically incorporated in new projects. Appraisal could be improved by providing a wider range of expertise, formulating expectations more realistically, and possibly by supplementing current methodologies.

Supervision would benefit from greater frequency and flexibility, better coverage of sociological aspects, and more emphasis on monitoring and evaluation.

Of continuing concern in settlement projects are the relatively high cost and poor cost recovery from beneficiaries. Measures identified to reduce costs include combining sponsored settlers with native and spontaneous settlers; greater private sector involvement; orienting settlements more towards regional roads and market towns; investing less in settler housing, and involving settlers more in project aspects. Better cost recovery could be achieved by making cost recovery policies explicit at the time of settler recruitment; through more efficient collection; using a portion of collected funds to benefit settlers directly; and by establishing group liability through settler organizations.

Another recommendation is that the Bank should clarify and formalize its policy on land settlement. Furthermore, Bank staff should become better aware of settlement issues, with more attention paid to beneficiaries and farming/production systems. Small-scale settlements should be linked to adjacent communities. Assistance should be extended on a pilot basis to non-farm enterprise development and employment generation. Bank funding should be provided for all key project components, including urban development. The design, management and organization of settlement schemes should take due account of aspects of transferring responsibilities at project completion to suitable local organizations. In cases of poor performance, the Bank should promptly initiate remedial measures and apply sanctions as appropriate if performance remains unsatisfactory.

CASE STUDY, KHASHM EL GIRBA IRRIGATION SCHEME, SUDAN.

The Khashm el Girba project in eastern Sudan, west of the Atbara River, is part of the resettlement program for the High Aswan Dam project. It involves irrigation of about 200,000 ha of land by gravity flow from a storage reservoir on the Atbara River. The project began operation in 1964 (Abu Sin, 1985).

Because provision of irrigation water is frequently one of the major purposes of water resource development projects in developing areas and resettlement is one of the major problems associated with construction of storage reservoirs, study of the formulation of this project and the ensuing problems can be instructive for all water resource planners.

Sin (1985) postulates that the root of the problems is the difference between how planners and settlers define "development". The conflict of interest between management and settlers is one of the major causes of decline in productivity of the project and of other similar major agricultural development projects in the semi-arid areas of Sudan.

Population of the project area in 1980 was about 350,000, including some 150,000 tenants and their families and 200,000 nomads. About 30 percent of the tenants are Nubians. Butana nomads constitute over 60 percent of total tenants and 80 percent of nomadic tenants. The Nubians, who were relocated from the Aswan reservoir area, live in 25 villages with all services. The nomads, whose grazing lands were taken for the Khashm el Girba resettlement project, live in 52 villages poorly equipped with services.

The three main crops are cotton, wheat, and groundnuts.

Annual rainfall in the area ranges from 250 to 300 mm. Losses from evaporation and seepage are about 17 percent, and field application losses are about 14 percent, so that less than 70 percent of available water is actually used for irrigation.

In addition, the water supply reservoir is filling with sediment, reducing water supply for the project. The project was designed for an annual water supply of 1,620 million cubic metres. However, available water is now only about 775 million cubic metres per year, of which 650 million cubic metres is reserved for a sugar plantation added to the project at a later stage, leaving only 125 millions cubic metres for use by tenants. At the present rate of reservoir sedimentation, reservoir yield will be reduced to 500 million cubic metres per year by 1997, creating a serious water shortage.

Project objectives were:

1. To resettle 52,000 Nubians whose land was submerged by the High Aswan Dam project.

2. To provide tenancies for Butana nomads whose grazing lands were taken for the Khashm el Girba resettlement project.

Tenancies for the Butana nomads was part of the general policy of the Sudan government for sedentarizing nomads, and both objectives fit into the national policy for expansion of modern agriculture. Project output of cash crops of cotton and groundnuts would improve Sudan's balance of payments problem by increasing production of crops for export, and production of sugar and wheat would reduce agricultural imports. Almost

TABLE 3.4 Reasons for Satisfactory Performance of Agricultural Projects, 1985 and 1984/a

Contributory Factor	Percentage of Projects Affected by This Factor		Percentage of Projects Where This Factor Was					
			The Most Important		Second Most Important		Third Most Important	
	1985	1984	1985	1984	1985	1984	1985	1984
Design Merits:								
Appropriate Project Content (simplicity, sufficient local resources, or suitable technology	73	54	30	29	8	17	24	6
Appropriate Institutional Arrangements	65	51	14	9	27	29	11	9
Strong Borrower support (for project goals and, during implementation, for providing adequate local finance, input supplies)	73	57	30	26	22	14	3	11
Successful Procurement	8	6	5	-	8	-	-	-
Successful Execution of Civil Works	35	23	5	6	3	6	8	3
Good Institutional Performance	51	37	5	11	16	6	16	11
Good Performance by Consultants or Technical Assistance	14	14	-	3	5	3	3	3
Favourable Economic Conditions	8	17	3	3	3	6	-	3
Favourable Support of Pricing and Other Government Policies	11	6	3	3	-	-	-	-

/a The figures relate to 37 agricultural projects reviewed in 1985 and 34 in 1984.
After The World Bank, 1987.

TABLE 3.5 Reasons for Unsatisfactory Performance of Agricultural Projects, 1985 and 1979-84/a

Contributory Factor	Percentage of Projects Affected by This Problem		Percentage of Projects Where This Problem Was					
			The Most Important		Second Most Important		Third Most Important	
	1985	1979-84	1985	1979-84	1985	1979-84	1985	1979-84
Design Problems:								
Inappropriate Project Content (too complex, insufficient local resources, or unsuitable technology)	100	86	50	41	22	18	11	14
Inappropriate Institutional Arrangements	89	86	28	20	17	28	22	15
Insufficient Borrower support (for project goals and, during implementation, providing inadequate local finance, input supplies)	50	69	6	22	22	8	6	11
Problems with Procurement	6	34	-	-	-	4	-	8
Difficulty in Executing Civil Works	11	22	-	-	-	1	-	-
Poor Institutional Performance	50	51	6	3	11	8	6	15
Poor Performance by Consultants or Technical Assistance	11	22	-	-	6	4	-	1
Adverse Economic Conditions	67	61	11	5	17	9	17	11
Political Difficulties	17	31	-	7	-	4	11	3
Natural Calamities	-	12	-	-	-	3	-	1
Adverse Effect of Pricing and Other Government Policies	28	54	-	3	-	11	-	8

/a The figures relate to 18 projects reviewed in 1985 and 74 in 1979-84.
   After The World Bank, 1987.

all agricultural operations at the project are tightly controlled, and the tenants have no choice in crops and so on.

In the first five years of project operation, the project appeared to be successful, but in recent years problems of management, water shortages, lack of spare parts, etc. have caused a decline in productivity and a decrease in area cultivated. Animal husbandry dominated the area prior to implementation of the irrigation project, and with the decline in yields from cash crops, the nomads are turning again to pastoralism.

Formulation of the irrigation project not only neglected the settlers perceptions of development and change, but also appear to have neglected such factors as declining fertility, weeds, and water availability.

Monetary returns from agriculture in 1980 were about the same as the foreign exchange settlers realized from livestock sales. The nomads would prefer to return to a livestock economy. The Nubians would prefer to practice unregulated small-scale agriculture as they did in their original home area. At this point all tenants, both the Nubians and the nomads, agree that the irrigation project should be modified, but there is no general agreement as to how it should be restructured. However, the settlers experience and current perceptions indicate there is a greater potential for compromise than when they first settled.

REFERENCE

Agency for International Development (AID), 1971.

Abu Sin, M.E. 1985. Planners' and Participants' Perceptions of Development in the Semi-arid Lands of Sudan: A Case Study of the Khashm el Girba Scheme, in Natural Resources and Rural Development in Arid Lands: Case Studies from Sudan, H.R.J. Davies, ed., The United Nations University.

Baum, W.A. and Tolbert, S.M., 1985. Investing in Developing, The World Bank, Oxford University Press.

Bembridge, T.J., 1984. Aspects of Agriculture and Rural Poverty in Transkei, Second Carnegie Inquiry into Poverty and Development in South Africa, Cape Town.

Dantzig, G. B., 1963. Linear Programming and Extensions, Princeton Univ. Press.

Development Bank of Southern Africa, 1987. Transkei Development and Information.

Goodman, A.S., 1984. Principles of Water Resources Planning, Prentice-Hall Inc., New Jersey.

Hollingworth, B.E. 1988. Course on Water Resources in Developing Areas, University of the Witwatersrand, Johannesburg.

IWES, 1983. Water Supply and Sanitation in Developing Areas.

Jalee. P., 1969. The Third World in World Economy, Monthly Review Press, New York.

James, L.D. and Lee, R.R., 1971. Economics of Water Resources Planning,. McGraw-Hill Inc., New York.

King, J.A., 1967. Economic Development Projects and Their Appraisal, 1967. John Hopkins Press, Baltimore.

The World Bank, 1987. Operations Evaluation Department, The Twelfth Annual Review of Project Peformance Results.

Stone. A., 1984. A case study of water resources and water quality of Chalumna/Hamburg area of Ciskei, CCP 148.

Sutherland, F., 1988. Course on Water Resources in Developing Areas, University of the Witwatersrand, Johannesburg.

United Nations, 1968. Planning Water Resources Development, United Nations Office of Technical Co-Operation.

World Bank Report, 1980. Poverty and Human Development, Oxford University Press, New York.

World Bank, Operations Evaluation Department, 1987. The Twelfth Annual Review of Project Performance Results.

CHAPTER 4

SYSTEMS ANALYSIS AND OPTIMIZATION

INTRODUCTION

Planning for water resources development involves seeking optimum designs. The most efficient design of a structure is usually that which achieves the objectives in the most economic manner. In water resources planning, on the other hand, the cheapest design is not always the optimum, since other factors, such as provision of employment opportunity, encouragement of industries, and environmental and social impacts must be considered.

The efficiency of a system design can be evaluated in terms of an objective function which is a mathematical expression of the sum of the net benefits stemming from a project together with weighted non-economic factors. An approach to water resources systems design using mathematical techniques is described in this chapter.

In a mathematical model of a water resource system, relationships between variables are expressed as mathematical equations or inequalities, and the objective function is expressed as an algebraic function of these variables. Techniques for analysing such models are described.

Systems optimization usually involves benefit-cost comparisons. The benefits of using water for irrigation, urban requirements, hydro-electric power, recreation, health, navigation, water quality control and flood control should all be considered. Benefits may be divided into direct benefits, or the value of immediate products or services resulting from a project, and indirect benefits, which include all other benefits expressible in monetary terms. Intangible benefits are those which cannot be expressed in monetary terms. To achieve the optimum net benefit, a project design, for example a combination of dams, conduits and irrigation areas, is modified until the net benefit is a maximum (see Eckstein, 1961).

Under normal circumstances the design which yields the maximum difference between benefits and costs is the optimum. The benefit/cost ratio, however, may be employed to rank projects in order of priority.

In order to compare benefits and costs, costs are discounted to a common time period – either present value or average annual cash flows. The principles behind valuations are summarized in "An Introduction to Engineering Economics" published by the Institution of Civil Engineers (1969).

A criterion sometimes used to decide whether to embark on a project is the comparison of the rate of return with the interest rate. However this criterion is cumbersome in systems design as it does not directly lead to an optimum combination of variables.

The interest rate applicable in public projects is open to question. While private concerns will apply the rate at which they can borrow money, national projects involve factors such as the social rate of time preference and national economic growth rate.

If a limit is imposed on expenditure, it is necessary to optimize the design within this budget rather than eliminate sections of the project at a later stage.

Analysis should also account for the effects of price inflation with time and, in state planning, price subsidies and foreign exchange should be considered.

SYSTEMS ANALYSIS TECHNIQUES

The following sections explain some techniques of systems analysis, or operations research, which are useful in analysing water resources systems. It is only within the last decades that the methods have been applied to any great extent in the field of water resources, although they have long been used in other spheres. There is still not widespread use of the techniques in practice (IAHS, 1989) although simulation has gained acceptance, perhaps because it is simpler and more comprehensive generally for complex problems.

Queuing theory and dynamic programming have become routine techniques in inventory control, and mathematical optimization is used extensively in business management. Linear programming has been used to optimize industrial systems and even military manoeuvres, while transportation programming, as its name implies, yields least-cost systems for transporting goods between sources and demand centres. The most versatile tool is probably system simulation by computer. The application of this technique to water resources systems, which involves iterative application of the hydrological equation, is adequately desribed in text books. Whereas simulation is useful for analysis of pre-defined systems, the methods described here are essentially direct optimization techniques, i.e. they lead automatically to an optimum design. Principles behind the techniques are not verified rigorously; instead, the methods are demonstrated with the aid of simple examples, and the reasons for steps proved informally. Books such as Dorfman et al (1958), show applications of the methods in practice and Maass et al (1962) were the first to write an application in water resources.

LINEAR PROGRAMMING BY THE SIMPLEX METHOD

The Simplex method of linear programming is one of the most powerful techniques for optimizing linear systems. If a system can be defined by a set of linear equations or in-equalities and an objective expressed as a linear function of the variables, then there exists a direct method of reaching an optimum combination of the variables, (Loomba, 1964).

A two dimensional example (i.e. two variables) will illustrate why a linear system, as opposed to a non-linear system, can be solved by a mechanical process. Suppose a combination of two variables, X and Y, is to satisfy the following constraints:

$$Y \leq 2,$$
$$Y+2X \leq 3$$
$$X,Y, \geq 0$$

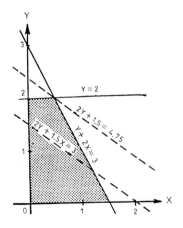

Fig. 4.1 Two Dimensional example

X and Y could represent two possible forms of development, e.g. irrigation and industrial, receiving water from a reservoir. It is desired to maximise the function 2Y+1.5Xm i.e., two units of Y are worth 1.5 units of X. The permissible domain for the variables is indicated by the shaded area in Figure 4.1. Any point in that domain satisfied the constraints, and the boundaries are the constraints with the inequality signs replaced by equal signs. Various values of the objective function are plotted on the same figure. The maximum value of 2Y+1.5X occurs at the intersection of the lines Y=2 and Y+2X=3. The value of the objectives function at any other point in the domain is less than the value corresponding to this point.

By comparing the values of the objective function at neighbouring junctions it will be found that its value always increases around the boundary in the direction of the optimum. By proceeding between successive intersections the optimum must eventually be reached, and the objective function will always increase each step until the optimum is reduced.

The domain of a linear system must always be convex, i.e. bulging outwards at each junction. (It is impossible to construct a junction which points inwards, since the bounding line would cross the domain). Therefore there are no local maxima at which the objective function ceases to increase when proceeding around the boundary, and once a point is found at which the objective function ceases to increase, that must be the true optimum. This would not necessarily be so if the boundaries were non-linear since a local maximum could exist between two intersections. The principles could be extended to more than two dimensions but would be difficult to illustrate graphically.

The Simplex Method – An Example

The Simplex method of optimization will be demonstrated with an example of a reservoir serving an irrigation area. The yield from storage for a certain recurrence interval of shortfall is indicated by the draft-storage curve, Figure 4.2. The maximum possible draft is limited by the extent of the irrigable area, to 1.78 thousand million cubic metres (TMC) per annum. The draft-storage curve is approximated by two straight lines, $q = 0.206 + 1.74s$ and $q=0.47+0.62s$ where s is the storage capacity in TMC and q is the draft in TMC per annum.

The constraints may be expressed algebraically as:

$q \leq 0.206+1.74s$

$q \leq 0.47+0.62s$

$q \leq 1.78$

Slack variables x, y and z are introduced into the inequalities to convert them to equations.

q−1.74s+x   = 0.206

q−0.62s   + y   = 0.47

q   + z   = 1.78

The variables q, s, x, y and z must all be positive.

    It is desired to maximize the net economic benefit of the scheme. The irrigation water has a net value, after subtracting farming costs and canal costs, of \$6 per annum per thousand cubic metres per annum, and the annual cost of the dam is \$2.5 per TMC capacity. The objective function can therefore be expressed as $6q−2.5s$

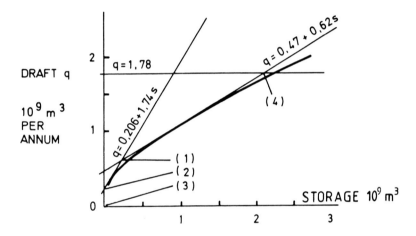

Fig. 4.2 Storage-draft curve and straight line approximation

    Table 4.1 illustrates the pattern in which the data are arranged for solution. The coefficients of the constraints form the main body of the matrix. Above each variable is its cost coefficient. To initialise the solution, the slack variables x, y and z are assigned the values 0.206, 0.47 and 1.78 respectively, to make the other variables equal zero. The three columns to the left of Table 4.1a indicate the current variables in the programme, their cost coefficients and their magnitudes.

    The numbers in any particular line of the main body of the matrix indicate the amount of the variable in the column labelled programme which would be displaced by one unit of the variable represented by the column of the number. Thus the 1 in the top line under the q column implies that one unit of q would displace one unit of x from the programme. Likewise by introducing 1 unit of s, it would be necessary to introduce an additional 0.62 units of y to keep the second constraint satisfied.

In order to determine whether it is worthwhile replacing the variables x, y and z in the programme by any of the other variables, a set of numbers (referred to as the net evaluation numbers or opportunity costs) is calculated. If one unit of q were introduced into the programme, then one unit of x, one unit of y and one unit of z would have to be dropped to satisfy the constraint equations. The objective function would increase by the amount $+6-(1\times0+1\times0+1\times0) = +6$. Likewise for each unit of s introduced into the programme it would be necessary to remove -1.74 units of x and -0.62 units of y (i.e. an increase of $+1.74$ units in x and $+0.62$ units in y) to keep the constraint equations satisfied. The net increase in benefit per unit of s introduced would be $-2.5-(-1.74\times0-0.62\times0-0\times0) = -2.5$. The benefit of bringing in one unit of any of the variables is indicated in the net evaluation row in Table 4.1a. To repeat, each opportunity cost is calculated by multiplying the entries in that column by the corresponding numbers in the profit column and subtracting the total from the objective coefficient of that column. If any of the net evaluation numbers is positive the indication is that the solution may be improved by introducing the column variable into the programme. Only one variable may be introduced into the programme at a time – usually that with the highest positive opportunity cost, or evaluation number. The column of the variable to be introduced is boxed in Table 4.1a. It is referred to as the key column.

There is a limit to the magnitude of the new variable, q, which may be introduced into the programme. If the first constraint is not to be violated, only 0.206 units of q may be introduced without causing x to turn negative. Similarly the second and third constraints would limit the amount to 0.47 and 1.78 respectively. The maximum amount which may be introduced is therefore 0.206, which is indicated by an asterisk to the right of Table 4.1a. The limiting constraint is referred to as the key row. The number which lies at the intersection of the key rown and key column is the key number.

After introducing the new variable the coefficients of the matrix have to be altered so that the relationships among the rates of replacement with the programme variables remain correct. The quantity of q in the new programme may not be the same as the quantity of x in the original programme at the top line of Table 4.1a. It is necessary to divide each of the numbers in the key row by the key number (in this case unity) to transform the row. The top line of Table 4.1b is the first constraint with each of the numbers divided by the key number so that the coefficient of the programme variable is unity.

The introduction of 0.206 units of q will cause the quantity of the other variables in the programme, namely y and z, to change. According to Table 4.1a, since one unit of q displaces one unit of y, the new value of y becomes 0.47-0.206=0.264. If the coefficient of q in the second row of the key column of Table 4.1a had been 2 instead of 1, it would have implied that one unit of q would displace 2 units of y, and the quantity of y in the programme would be reduced by (2/1)x0.206. To keep the constraint equations correct, each of the numbers in line 2 of Table 4.1a has subtracted from it the corresponding number in line 1. In general, the transformation rule for non-key rows is: Subtract from each number in a non-key row the corresponding number in the key row multiplied by the ratio formed by dividing the old row number in the key column by the key number. The matrix is thereby inverted to eliminate the key variable.

The process of studying the opportunity costs of each variable, introducing that variable with the greatest opportunity value into the programme, then inverting the matrix, is repeated until no further positive opportunity exists. In Table 4.1b the opportunity costs are indicated in the bottom row. The Opportunity for each column is calculated by multiplying the numbers in that column by the numbers in the profit column and subtracting the total from the objective coefficient. The largest positive opportunity cost, 7.9, indicates the key column. The maximum amount of s which may be introduced into the program is determined by comparing the replacement ratios for each row. Introducing 1 unit of s would mean displacing 1.12 units of y. Since there are only 0.264 units of y in the programme, the maximum amount of s which could be introduced without violating the second constraint would be 0.264/1.12=0.235. Row 1 indicates an irrelevant quantity since the introduction of s only increases q and does not violate any non-negotiating constraints. Since 0.235 is the maximum amount which may be introduced, the second row becomes the key row. The martrix is transformed according to the above rules, into the new matrix Table 4.1c.

Summary of Simplex Method:

1.   Write down objective function
2.   Set up linear constraints

TABLE 4.1a

Programme variable	Objective coefficient:		6	-2.5	0	0	0	
	Pro-fit	Quan-tity	q	s	x	y	z	Replacement ratio
x	0	0.206	1	-1.74	1	0	0	0.206/1=0.206*
y	0	0.47	1	-0.62	0	0	0	0.47/1=0.47
z	0	1.78	1	0	0	0	1	1.78/1=1.78
Net evaluation row			6*	-2.5	0	0	0	

TABLE 4.1b

Programme variable	Pro-fit	Quan-tity	6	-2.5	0	0	0	
			q	s	x	y	z	
q	6	0.206	1	-1.74	1	0	0	0.206/-1.74=-0.119
y	0	0.264	0	1.12	-1	1	0	0.264/1.12=0.235*
z	0	1.574	0	1.74	-1	0	1	1.574/1.74=0.905
			0	7.9*	-6	0	0	

TABLE 4.1c

Programme variable	Pro-fit	Quan-tity	6	-2.5	0	0	0	
			q	s	x	y	z	
q	6	0.616	1	0	-0.55	1.55	0	0.616/-0.55=-1.12
s	-2.5	0.235	0	1	-0.895	0.895	0	0.235/-0.895=-0.264
z	0	1.164	0	0	0.55	-1.55	1	1.164/0.55=2.12*
			0	0	1.1*	-7.1	0	

TABLE 4.1d

Optimum programme

Programme variable	Pro-fit	Quan-tity	6	-2.5	0	0	0
			q	s	x	y	z
q	6	1.78	1	0	0	0	1
s	-2.5	2.13	0	1	0	-1.63	1.63
x	0	2.12	0	0	1	-2.82	1.82
			0	0	0	-4.07	-1.93

3.   Convert to equations by adding dummy variables

4.   Write coefficients in table form

5.   Calculate evaluation numbers for each variable

6.   Select variable with maximum (or minimum) evaluation number

7.   Find maximum amount of key variable from replacement ratios

8.   Invert matrix to eliminate previous variable

9.   Repeat steps 5-8 until no more improvement possible

10. Write down remaining variable values.

Other Cases (Dantzig, 1963)

The foregoing example is a case of maximizing an objective function, with constraints expressed as inequalities. Cases may occur with equalities, or it may be desired to minimize a function. Following is a summary of the initial procedures for reducing all cases to the form of the above example.

(i)   It is desired to maximize the function $300a+180b$

Constraints are $8a+5b \leq 80$

$4a+2b \leq 70$

$a, \ b \ \geq \ 0.$

The Simplex form of the equation is

$8a+5b+x \ = \ 80$

$4a+2b+y \ = \ 70$

where x and y are positive slack variables.

Objective function : Maximize $300a+180b+0x+0y$.

Procedure then as in above example.

(ii) Maximize $300a+180b$

Constraints $8a+5b \ = \ 80$

$4a+2b \ = \ 70$

$a,b \ \geq \ 0$

The Simplex form of the equation is

$8a+5b+u \ = \ 80$

$4a+2b+v \ = \ 70$

where u and v are positive artificial slack variables.

Objective function : 300a+180b-Mu-Mv

where M is a very large positive number

Procedure then as before.

(iii) Minimize   300a+180b

     Constraints   8a+5b = 80

                      4a+2b = 70

                       a,b $\geq$ 0

The Simplex form of the equation is

                 8a+5b+u = 80

                 4a+2b+v = 70

where u and v are positive artificial slack variables.

Objective function : Maximize – 300a–180b–Mu–Mv

where M is any very large positive number.

Procedure then as before.

(iv) Minimize   300a+180b

     Constraints   8a+5b $\geq$ 80

                      4a+2b $\geq$ 70

                       a,b $\geq$ 0

The Simplex form of the equation is

                 8a+5b–x+u = 80

                 4a+2b–y+v = 70

where x and y are positive slack variables and u and v are positive artificial slack variables.

Objective function : Maximize –300a–180b+0x+0y–Mu–Mv.

where M is any very large positive number.

Procedure then as before.

Notes: The objective coefficients of a and b could be positive or negative in the above cases.

     The coefficients of a and b in the constraint equations could be positive or negative.

     The constants on the right hand side of the constraints must be positive. If they are negative, multiply through by –1 and invert inequality signs.

     The value of M must be sufficiently large to ensure the variables u and v are zero in the final solution.

     Combinations of less than, and greater than inequalities, and equal signs may be treated by considering each constraint individually.

An alternative method to that in cases (iii) and (iv) above exists for solving the minimization case.

Instead of multiplying the objective coefficients by minus one to convert to a maximization case, the column with the lowest negative opportunity cost could be introduced into the programme. For the maximization case it will be recalled, the column with the highest positive opportunity cost was brought into the programme.

SHADOW PRICES AND THE DUAL FUNCTION

Returning to the reservoir problem the numbers in the net evaluation row of the final matrix have an interesting interpretation. The opportunity value of y is -4.07. In other words, by allowing one unit of slack, y, in the solution, the net benefit from the system would decrease by 4.07. Similarly, by introducing one unit of z, or decreasing the irrigation supply by 1 TMC per annum the net benefit would decrease by 1.93 units. On the other hand, if there were more irrigable land available, the net benefit could be increased by 1.93 units per additional TMC per annum. The numbers 4.07 and 1.93 are referred to as the 'shadow prices' of y and z. The value of the resources may be computed from these numbers and from the quantity of available resources. Thus, $4.07 \times 0.47 + 1.93 \times 1.78 = 5.4$, which is the same as the net benefit of the system derived earlier. The value of the resources could also have been computed from the dual linear programming problem:

Recall the original problem:

$q - 1.74s \leq 0.026$

$q - 0.62s \leq 0.47$

$q \qquad \leq 1.78$

Minimize $6q - 2.5s$

The dual problem is:

$a + b + c \geq 6$

$-1.74a - 0.62b \geq -2.5$

Minimize $0.206a + 0.47b + 1.78c$

It will be observed that the matrix for the dual is the matrix of the original problem turned on its side. a, b, and c are the worths, or shadow prices of each of the original constraints.

Solution of the dual will yield:

$b = 4.07$ $\qquad c = 1.93$ $\qquad$ which coincide with the shadow prices of y and z discussed above.

Degeneracy

It may happen at any stage during the computations that two replacement ratios are equal. In such case, whichever of the two rows is selected as the key row, the number in the quantity column in the other row will become zero when the new matrix is formed. When the replacement ratios for the new matrix are calculated, it will be observed that the ratio in the problem row will be zero. In such cases, the ratio is merely assumed to have a very small positive value, and the computations proceed as before.

COMPUTER PROGRAM

To optimize a reasonably large matrix by hand using the Simplex procedure would prove tiresome. Fortunately the technique is ideally suited to solution by computer. Accompanying is a FORTRAN computer program for optimizing by the Simplex method. The symbols which require explanation are tabulated.

A   –   Objective coefficient of a variable in the programme
B   –   Objective coefficient
BEN –   Net benefit
E   –   B–SXA
E   –   Maximum E
I   –   Sequential number of columns. Varies from 0 to NM
          reading from left to right.
IM   –   I corresponding to RM (key row No.)
IV   –   Sequential I – No. of a variables in the programme
J   –   Sequential No. of rows. Varies from 0 to M reading
          from top to bottom
JM   –   J corresponding to EM (key column No.)
N   –   No. of variables
NM   –   N+M, or total No. of columns
NW   –   N+1
M   –   No. of constraints or rows
R   –   Replacement ratio
RM   –   Minimum positive R
SXA –   Sum of XA's
X   –   Matrix coefficient
XA   –   X×A

COMPUTER PROGRAM

```
C       LINEAR PROGRAMMING
        READ 1,N,M
1       FORMAT (2I4)
        NM=N+M
        NW=N+1
        DIMENSION  X(49,21),XS(49,21)
        DIMENSION IV(21),A(21),Z(21),ZS(21), R(21), XA(21)
        DIMENSION E(49), B(49)
         READ 2, (B(J), J=1,N)
2       FORMAT (11F7.0)
        READ 3,(Z(I), I=1,M)
         READ 4, ( (X(J,I),J=1,N),I=1,M)
3       FORMAT (11F4.1)
4       FORMAT (11F7.3)
         DO  48 J=NW,NM
48      B(J) = -99999999.
        DO 20 I=1,M
        DO 19 J=NW,NM
19      X(J,I)=O.
        X(N+I,I)=1.
        IV(I)=N+I
20      A(I)=B(N+I)
18      DO 21 J=1,NM
        SXA=O.
        DO 22 I=1,M
        XA(I)=X(J,I) *A(I)
22      SXA=SXA+XA(I)
21      E(J)=B(J)-SXA
        EM=O.
26      DO 24 J=1,NM
        IF (E(J)-EM)24,24,25
25      EM=E(J)
        JM=J
        GO TO 26
24      CONTINUE
        IF(EM)27,27,28
28      DO 32 I=1,M
        IF(Z(I))30,31,30
31      Z(I)=.00001
30      IF(X(JM,I))32,33,32
33      X(JM,I) =.000000001
32      R(I)=Z(I)/X(JM,I)
        RM=1000000000000.
34      DO 36 I=1,M
        IF(R(I))36,36,37
37      IF (R(I)-RM)38,36,36
38      RM=R(I)
        IM=I
        GO TO 34
36      CONTINUE
        IF (RM-1000000000.) 40,39,39
```

```
39      PRINT 5
5       FORMAT(17HRM=1000000000000.)
        GO TO 27
40      DO 42 I=I,M
        ZS(I)=Z(I)
        DO 42 J=1,NM
42      XS (J,I)=X(J,I)
        DO 43 I=1,M
        Z(I)=ZS(I)-ZS(IM)*XS(JM,I)/XS(JM,IM)
        DO 43 J=1,NM
43      X(J,I)=XS(J,I)-XS(J,IM)*XS(JM,I)/XS(JM,IM)
        Z(IM)=ZS(IM)/XS(JM,IM)
        DO 44 J=1,NM
44      X(J,IM)=XS(J,IM)/XS(JM,IM)
        A(IM)=B(JM)
        IV(IM)=JM
        GOTO 18
27      BEN=O
        DO 45 I=1,M
45      BEN=BEN+A(I)*Z(I)
        PRINT 6
6       FORMAT (33H    IV       A         Z             R)
        DO 46 I=1,M
46      PRINT 7, IV(I),A(I),Z(I),R(I)
7       FORMAT (14,1X,E10.4, 1X,E10.4,1X,E10.4)
        DO 47 I=1,M
47      PRINT 8, (X(J,I), J=1,NM)
        PRINT 9, BEN
        PRINT 8, (E(J),J=1,NM)
9       FORMAT (9H BENEFIT =, E10.4)
8       FORMAT (5(10(1X,F8.3)/))
        END
```

XS   –   Matrix coefficient

Z   –   Quantity of a programme variable

ZS   –   Quantity of a programme variable

The FORTRAN statements up to number 48 are concerned with the input data and can be varied to suit the problem. The programme as it stands can optimize a problem having 21 constraints and 49 variables including slack and artificial slack variables. The values of N and M, the objective coefficients, constants and matrix coefficients are punched on cards to the format indicated in the programme.

After computations the machine will print the programme variables, their objective coefficients and optimum magnitudes. Also printed are the final replacement ratios, the final Simplex matrix, the net benefit, and the net evaluation row.

MULTI-OBJECTIVE PLANNING

It is recognized that maximization of net economic benefit is not the only objective in most development programs. The methods of handling the other objectives vary. The shadow pricing of non-economic factors is one way (see Fig. 4.3), another is to define non-economic objectives and perform multi-objective analysis to achieve an optimum compromise (Major, 1977).

Objectives in water resources planning in developing areas could include: Maximization of cash injection-national or regional

       Maximization of benefit – cost difference

       Maximization of benefit – cost ratio with expenditure budget
         limits

       Redistribution of wealth

       Environmental quality, including protection of resources

       Social upliftment

       Health improvement

       Provision of employment to improve stability, recirculate wealth,
        or provide training.

       Regional objectives can include in addition to above

       Development of selected reserves

       Attraction of further industry

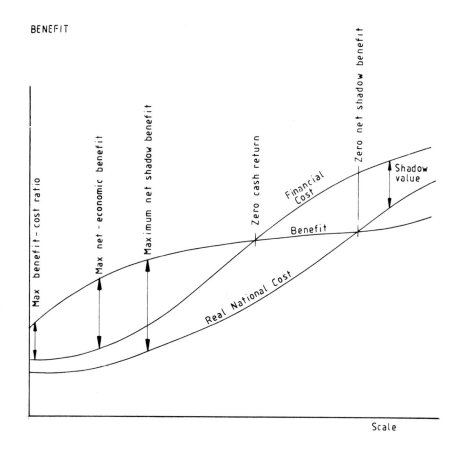

BENEFIT

Max benefit - cost ratio

Max net - economic benefit

Maximum net shadow benefit

Zero cash return

Zero net shadow benefit

Financial Cost

Benefit

Real National Cost

Shadow value

Scale

Fig. 4.3 The effect of shadow prices on optimum project extent

National and regional objectives may conflict. For example investment in particularly backward areas may cause a net drain on national resources for many years. This could set back national development particularly for small and sensitive countries.

The way suggested by Major (1977) to achieve a balanced optimum between national and regional objectives is to plot the benefits on a graph of regional versus national benefit. The line representing equal marginal benefits to both objectives is called the net benefit transformation curve.

On the same graph are plotted times of equal social preference (or any other objective). The optimum combination of objectives can then usually be identified by inspection.

REFERENCES

Dantzig, G.B., 1963. Linear Programming and extensions. Princeton Univ. Press.

Dorfman, R., Samuelson, P.A. and Solow, R.M., 1958. Linear programming and economic analysis. McGraw Hill, N.Y.

Eckstein, O., 1961. Water Resources Development, The Economics of Project Evaluation, Harvard Univ. Press, Cambridge.

Institution of Civil Engrs., 1969. An Introduction to Engineering Economics, London.

International Association of Hydrological Sciences, 1989. From Theory to Practice. Proc. Workshop on Systems Analysis in Water Resources Planning. Baltimore.

Loomba, N.P., 1964. Linear Programming. McGraw Hill,, N.Y.

Maass, A., Hufschmidt, M.M., Dorfman, R. Thomas, H.A., Marglin, S.A. and Fair, G.M. 1962. Design of Water Resource Systems. Macmillan, London.

Major, D.C., 1977. Multi Objective Water Resource planning. Amer. Geophys. Union Water Ress. Monograph 4, Washington, 81p.

CHAPTER 5

DECOMPOSITION OF COMPLEX SYSTEMS

EXAMPLE INVOLVING IRRIGATION

   Theoretically the linear programming process could be applied to the optimization of any linear system. However in some cases the problem could become unwieldy if treated using normal linear programming techniques. Very large problems exceed the capacity of available computers, and it may be desirable to sub-divide the problem into reasonably sized components. Some of the constraints could be assembled into a transportation programme which is generally most economically solved by hand calculation. A number of sub-programmes could be set up. These would be linked by a master programme. The penalty for such simplification is that the master programme and sub-programmes will have to be solved successively a number of times (Dantzig, 1963).

   The method will be illustrated by an example involving one transportation sub-programmme and a set of master constraints. The example involves optimizing an irrigation plan. To follow the reasoning demands a thorough knowleddge of linear programming and dual functions..

   There are two irrigable areas, referred to as K and L, and two possible sources of water, A and B. Reservoir A can supply 300 Mℓ/d (megalitres a day) and reservoir B can supply 700 Mℓ/d. There is a tract of 14600 hectares (ha) of arable land at locality K which needs 1 m of water per annum per ha and another 12166 ha at locality L which needs 1.5 m of water per annum per ha. Thus the maximum requirement of area K is $14600 \times 10000 \times 1/365 \times 1000 = 400$ Mℓ/d, and of area L, $12166 \times 10000 \times 1.5/365 \times 1000 = 500$ Mℓ/d. The cost of water conveyance to either area from either source is indicated in million dollars per annum per 100 Mℓ/d, at the top of the relevant grid positions in Table 5.1. The objective is to minimize the total conveyance cost.

   In Table 5.1 each column represents a demand and each row source of water. A slack column takes up the surplus water from the dams. An artificial slack row is also included to allow for each which could possibly lie unirrigated. The amount of 'water' assigned to this row is sufficiently large to permit all the area to remain unirrigated.

   The problem as so far outlined is a simple transportation programming problem. However, additional constraints which cannot be incorporated into the transportation sub-programme, are imposed for various reasons. In order to ensure balanced agricultural development in relation to the remainder of the economy, it is necessary that the area to be irrigated total 23000 ha. As the available irrigable area totals 26700 ha not all the area need be irrigated to meet this requirement, so slack, S, is introduced into each of the columns in Table 5.1. Stated algebraically in terms of variables S_K and S_L this is $(400-S_K)$ $36.5/1+(500-S_L)$ $36.5/1 = 23000$.

or $L.5S_K+S_L = 155$.

Also prior to expansion of the project, there existed 10950 ha in area K which was under irrigation. For political reasons it is necessary that this area shall not be abandoned. Thus a second constraint becomes $(400-S_K)36.5/1 \geq 10950$

or $S_K \leq 100$.

In Simplex form this becomes $S_K+x = 100$ where x is a slack variable with zero value.

The complete problem incorporates two parts: it would appear simpler to solve these separately rather than to attempt to solve one giant linear programming problem. The transportation programme may be referred to as a sub-programme, and the set of other constraints as the master programme. The method of solution is to assign artificial costs to the slack in columns K and L of Table 5.1 in order to control the quantity of water made available for irrigation. If zero cost were attached to the column slacks, as much as possible of the water would be allocated to those cells during the transportation programming process, whereas if too large a cost were assigned, the quantities in those cells would decrease and as a result the master constraints might be violated. The optimum premium cost to be assigned to those cells is determined from the dual to the master programme; this yields the shadow price of the master constraints. The process of optimization involves interating successive sub-programmes and master programmes until there is no further improvement in the objective function. Each sub-programme solution yields a new variable to be introduced into the master programme, and the premium costs are then revised.

Suppose that a number of feasible but not necessarily optimal solutions to the sub-programmes were at hand. Corresponding to solutions 1 and 2 say, there are transportation sub-programmes 1 and 2. Provided the allocations in each of the transportation programmes satisfy the demand and availability constraints, then any weighted combination of them will also satisfy the constraints. Thus if each entry in transportation sub-programme 1 is multiplied by 1/3 and added to 2/3 the corresponding entry in sub-programme 2, the result will still satisfy the constraints. This suggests that each sub-programme could be represented by a weight in the master programme and the sub-programme constraints could be replaced by the single constraint that the sum of the weights should total unity. The master programme would then become:

$$(C_1)V_1 + (C_2)V_2 + (C_3)V_3 \qquad\qquad = z \text{ (objective function)}$$
$$(1.5S_K + S_L)_1 V_1 + (1.5S_K + S_L)_2 V_2 + (1.5S_K + S_L)_3 V_3 \quad = 155$$
$$(S_K)_1 V_1 + (S_K)_2 V_2 + (S_K)_3 V_3 + x \qquad = 100$$
$$V_1 \qquad\qquad +V_2 \qquad\qquad +V_3 \quad = 1$$

where the C's are total conveyance costs and the V's weights. Subscripts V_1, $+ V_2$ and $+ V_3$ refer to the sub-programmes 1, 2 and 3 respectively.

Computations are initialized by generating an arbitrary, not necessarily optimum, basic solution to the sub-programme. The initial assignments in Table 5.1 were calculated by proceeding from the top left to bottom right as outlined in the section on transportation programming.

TABLE 5.1 Sub-programme 1.

Area	K	L	Slack	Yield Mℓ/d	Cost	$1.5S_K$	$+S_L$	S_K
Source	1.2	0.6	0					
A	300			300	300×1.2 = 360	0	0	0
	1.5	1.0		0				
B	100	500	100	700	100×1.5 = 150			
				0				
Slack S			1000	1000	500×1.0 = 500			
					1000			

Limit Mℓ/d   400   500   1000

A linear programming problem is formulated with the initial sub-programme solution and the master constraints. The variables to be considered are the weights, V, by which the solution to the sub-programme is to be multiplied before it is included in the master problem.

The objective function is to minimize $1010V_1 \qquad = Z$
subject to the master constraints $(1.5S_K + S_L)_1 V_1 \qquad = 155$
$$(S_K)_1 V_1 \qquad \leq 100$$

In addition, the sum of the weights must total unity:

$$V_1 = 1$$

Expressed in Simplex form,

$$1010V_1 = Z \text{ (min)}$$
$$0V_1 + a = 155$$
$$0V_1 \quad + x = 100$$
$$V_1 \quad + b = 1$$

where a and b are artificial slack variables with very large cost coefficients called M. The master programme is optimized by linear programming. This is a minimization case, so the largest negative number in the net evaluation row indicates the variable to be brought into the programme.

TABLE 5.2a Master Programme 1

			1010	M	0	M	
Variable	Profit	Quantity	V_1	a	x	b	
a	M	155	0	1	0	0	155/0
x	0	100	0	0	1	0	100/0
b	M	1	1	0	0	1	1/1=1*
			1010	0	0	0	
			-M*				

TABLE 5.2b

			1010	M	0	M
Variable	Profit	Quantity	V_1	a	x	b
a	M	155	0	1	0	0
x	0	100	0	0	1	0
V_1	1010	1	1	0	0	1
			0	0	0	M-1010

The variables a, x and V_1 appear in the above optimum programme. To determine whether it is worthwhile to intoduce other sub-programmes into the master programme, the corresponding opportunity costs must be determined. It will be recalled that opportunity costs are calculated by comparing the actual price on any particular variable with the sum of the column coefficients multiplied by certain numbers which will be denoted

p_1, p_2 and p_3 and which correspond to rows 1, 2 and 3 in the programme. The values of the multipliers are not immediately apparent from the final tableau of the above linear programme, since the original constraint coefficients have been transformed. Values of the multipliers may be calculated by using the fact that opportunity costs of the optimum variables in the programme are zero. Opportunity costs corresponding to the variables V_1, a and x in Table 5.2a are calculated as follows:

$$1010 + 0p_1 + 0p_2 + 1p_3 = 0$$
$$M + 1p_1 + 0p_2 + 0p_3 = 0$$
$$0 + 1p_1 + 0p_2 + 0p_3 = 0$$

Coefficients in the above equations are the numbers in Table 5.2a under the variables V_1, a and x respectively. Solution of the above equations, which together in fact comprise the dual programme to the master programme, yields the multipliers p_1, p_2 and p_3 as follows:

$$p_1 = -M, \quad p_2 = 0, \quad p_3 = -1010$$

If another sub-programme represented by V_2 can be found which, when these multipliers are used, indicates an opportunity cost less than zero, then according to the Simplex criterion that sub-programme should be introduced into the master programmme. A second sub-programme will be represented by weight V_2, and the coefficients in the new column in the master programme will be:

$$(1.5S_K + S_L)_2$$
$$(S_K)_2$$
$$1$$

The opportunity cost of the new sub-programme will be the actual cost together with the products of the coefficients and the corresponding multipliers p_1, p_2 and p_3. To discern the sub-programme associated with minimum opportunity cost, the sub-programme is optimized with new prices. In addition to actual prices, premium prices are generated as follows:

Corresponding to row 1 : $(1.5\ S_K + S_L)_2 \times p_1$
Corresponding to row 2 : $(S_K)_2 \times p_2$
Corresponding to row 3 : $1 \times p_3$

Thus the prices placed on S_K and S_L for the new sub-programme become:

for S_K : $0 + 1.5_{p1} + 1_{p2} = -1.5M$
for S_L : $0 + 1_{p1} = -M$

These prices are used in optimizing sub-programme 2. Table 5.1 is taken as a starting array for the transportation programming procedure which is solved and the results are given in Table 5.3.

TABLE 5.3 Sub-programme 2.

	K	L		Cost	$1.5S_K+S_L$	S_K
	1.2	0.6	0			
A		300		0	1100	400
	1.5	1	0			
B		700				
	-1.5M	-M	0			
S	400	500	100			

In the master progamme the opportunity cost of sub-programme 2 is :

$$0 - 1.5M \times 400 - M \times 500 - 1010 = -1010 = -1100M - 1010,$$

which is negative, therefore it will be worthwhile to introduce the new programme into the master programme. The simplex equations for the new master programme are:

$$1010V_1 + 0V2 \qquad\qquad = Z$$
$$0V_1 + 1100V_2 + a \qquad = 155$$
$$0V_1 + 400V_2 + x \qquad = 100$$
$$V_1 + V_2 + b \qquad = 1$$

The master programme is optimized by linear programming to determine which variables occur in the programme.

TABLE 5.4a Master Programme 2

			1010	0	M	0	M	
Variable	Profit	Quantity	V1	V2	a	x	b	
a	M	155	0	1100	1	0	0	155/1100=0.141*
x	0	100	0	400	0	1	0	100/400
b	M	1	1	1	0	0	1	1/1
			1010	-1101M*	0	0	0	
			-M					

TABLE 5.4b

Variable	Profit	Quantity	1010 V₁	0 V₂	M a	0 x	M b	
V_2	0	0.141	0	1	0.001	0	0	0.141/0
x	0	43.5	0	0	-0.364	1	0	43.5/0
b	M	0.859	1	0	-0.001	0	1	0.859/1=0.859*
			1010	0	M-	0	0	
			-M*					

TABLE 5.4c

Variable	Profit	Quantity	1010 V₁	0 V₂	M a	0 x	M b
V_2	0	0.141	0	1	0.001	0	0
x	0	43.5	0	0	-0.364	1	0
V_1	1010	0.859	1	0	-0.001	0	1

Dual:

$$1010 + 0p_1 + 0p_2 + p_3 = 0 \quad \text{(coefficients of } V_1)$$
$$0 + 1100p_1 + 400p_2 + p_3 = 0 \quad \text{(coefficients of } V_2)$$
$$0 + 0p_1 + p_2 + 0p_3 = 0 \quad \text{(coefficients of } x)$$
$$p_1 = 0.918, \quad p_2 = 0, \quad p_3 = -1010$$

Prices:

S_K: $0 + 1.5 \times 0.918 + 1 \times 0 = 1.38$

S_L: $0 + 1 \times 0.918 = 0.92$

TABLE 5.5 Sub-programme 3

	K	L		Cost	$1.5S_K+S_L$	S_K
A	1.2	0.6 [300]	0	180	800	400
B	1.5	1	0 [700]			
	1.38 [400]	0.92 [200]	0 [400]			

Opportunity cost = 180+1.38×400+0.92×200−1010 =   −94

                  <0 so sub-programme 3 should be brought into the

master programme.

Master programme Simplex equations:

$$1010V1+ 0V2 \quad + 180V3 \qquad\qquad = Z$$
$$0V1+1100V_2 \quad + 800V3 + a \qquad = 155$$
$$0V1+ 400V2 \quad + 400V3 + x \qquad = 100$$
$$V_1 \quad +V_2 \quad + V_3 \qquad\qquad = 1$$

TABLE 5.6a Master programme 3

			1010	0	180	M	0	M	
Variable	Profit	Quantity	V_1	V_2	V_3	a	x	b	
a	M	155	0	1100	800	1	0	0	155/1100 = 0.141*
x	0	100	0	400	400	0	1	0	100/400
b	M	1	1	1	1	0	0	1	1/1
			1010	-110/M*	180	0	0	0	
			-M		-801				

TABLE 5.6b

Variable	Profit	Quantity	1010 V_1	0 V2	180 V_3	M a	0 x	M b	
V_2	0	0.141	0	1	0.725	0.001	0	0	0.141/0
x	0	43.5	0	0	109	-0.364	1	0	43/5
b	M	0.859	1	0	0.275	-0.001	0	1	0.859/1 = 0.859*
			1010 -M*	0	180 -0.275M	M-	0	0	

TABLE 5.6c

Variable	Profit	Quantity	1010 V_1	0 V_2	180 V_3	M a	0 x	M b	
V_2	0	0.141	0	1	0.725	0.001	0	0	0.141/0.735=0.193*
x	0	43.5	0	0	109	-0.364	1	0	43.5/109 = 0.4
V_1	1010	0.859	1	0	0.275	-0.001	0	1	0.859/0.275 = 3.1
			0	0	-98*	M-	0	M-	

TABLE 5.6d

Variable	Profit	Quantity	1010 V_1	0 V_2	180 V_3	M a	0 x	M b
V_3	180	0.194	0	1.38	1	0.001	0	0
x	0	22.5	0	-155	0	-0.51	1	0
V_1	1010	0.806	1	-0.38	0	-0.001	0	1
			0	136	0	M-	0	M-

The variable which was rejected from the master programme in Table 5.6 need not be re-considered

Dual:

$$1010 + 0p1 + 0p2 + p3 = 0$$
$$180 + 800p1 + 400p2 + p3 = 0$$
$$p2 = 0$$
$$p1 = 1.04, \ p2 = 0, \ p3 = -1010$$

Prices

$$S_K : 0 + 1.5 \times 1.04 + 1 \times 0 = 1.56$$
$$S_L : 0 + 1 \times 1.04 = 1.04$$

TABLE 5.7 Sub-programme 4

	K	L		Cost	$1.5S_K+S_L$	S_K
A	1.2	0.6 [300]	0	980	0	0
B	1.5 [400]	1 [200]	0 [100]			
S	1.56	1.04	0 [1000]			

Opportunity cost = $980 + 1.56 \times 0 + 1.04 - 1010 = -30$
                                       0

So sub-programme 4 should be brought into master.

Master programme simplex equations :

$$1010V_1 + 180V_3 + 980V_4 = Z$$
$$0V_1 + 800V_3 + 0V_4 + a = 155$$
$$0V_1 + 400V_3 + 0V_4 + x = 100$$
$$V_1 + V_3 + V_4 + b = 1$$

TABLE 5.8a Master programme 4.

			1010	180	980	M	0	M	
Variable	Profit	Quantity	V_1	V_3	V_4	a	x	b	
a	M	155	0	800	0	1	0	0	155/800 = 0.194
x	0	100	0	400	0	0	1	0	100/400
b	M	1	1	1	1	0	0	1	1/1
			1010	180	980	0	0	0	
			-M	-801M*	-M				

TABLE 5.8b

			1010	180	980	M	0	M	
Variable	Profit	Quantity	V_1	V_3	V_4	a	x	b	
V_3	180	0.194	0	1	0	0.001	0	0	0.194/0
x	0	22.5	0	0	0	-0.5	1	0	22.5/0
b	M	0.806	1	0	1	-0.001	0	1	0.806/1=0.806*
			1010	0	980	M-	0	0	
			-M		-M*				

TABLE 5.8c

			1010	180	980	M	0	M
Variable	Profit	Quantity	V_1	V_3	V_4	a	x	b
V_3	180	0.194	0	1	0	0.001	0	0
x	0	22.5	0	0	0	-0.5	1	0
V_4	980	0.806	1	0	1	-0.001	0	1
			30	0	0	M-	0	M-

Dual:

$$180 + 800p_1 + 400p_2 + p_3 = 0$$
$$980 + 0p_1 + 0p_2 + p_3 = 0$$
$$p_2 = 0$$
$$p_1 = 1, \quad p_2 = 0, \quad p_3 = 0$$

Prices

S_K: $0 + 1.5 \times 1 + 1 \times 0 = 1.5$

S_L: $0 + 1 \times 1 \qquad = 1$

TABLE 5.9 Sub-programme 5

	K	L		Cost	$\dfrac{1.5S_K+S_L}{}$	S_K
A	1.2	0.6	0	180	800	400
		300				
B	1.5	1	0			
			700			
	1.5	1	0			
	400	200	400			

Opportunity cost = $180 + 1.5 \times 400 + 1 \times 200 - 980 = 0$

i.e.   No further improvement

The optimum programme is therefore

$V_3 = 0.194$, $V_4 = 0.806$, $x = 22.5$.

Table 4.10 is a combination comprising 0.194 x sub-programme 3 plus 0.806 x sub-programme 4.

TABLE 5.10 Optimum programme

	K	L	Slack	Yield Mld
A	1.2	0.6	0	300
		300		
B	1.5	1	0	700
	322.5	161	216.5	
slack			0	
	77.5	39	883.5	
Limit				
Mld	400	500		

The principle of decomposition could be applied to a wide range of optimization problems. In some cases the full rigorous procedure in arriving at the solution could be shortened. If the above irrigation problem had not been subject to the additional constraint involving the minimum irrigable area at K, solution would have been considerably simpler. The master constraint would have comprised only two constraints, and it would have been possible to determine simply by inspection which of the sub-programmes to reject. Linear programming optimization of the master programme would thus have been unnecessary.

Although the sub-programmes in the above examples took the form of transportation programmes, they could just as well have been linear programmes or dynamic programmes. The successive steps, using premium costs added to actual costs for sub-programme solutions, would proceed in a similar fashion to those for the example solved.

REFERENCES

Dantzig, G.B., 1963. Linear programming and extensions. Princeton Univ. Press, Princeton. Ch. 23.

CHAPTER 6

A PLANNING MODEL (Stephenson, 1989)

INTRODUCTION

The changes and needs associated with rapid development can tax the water resources of a country. More and bigger water supply schemes may continuously have to be planned and implemented as the economy expands. If planning is to succeed in coping with the economic development, bold steps should be taken. Since we have a situation where relatively few skilled and professional people must plan and cater for a large population, it is apparent that planning and design of new projects will continue to lag unless far reaching methods are used. In fact, modern developments in systems analysis and increasing availability of computers provide the necessary tools. Not only will computers speed up analysis, but automatic data assembly and analysis by computer is becoming essential to cope with the vast volumes of data requiring analysis.

This chapter describes a method whereby plans for all levels of development could be compiled. A mathematical model representing the various sectors of the economy at different levels would be set up as illustrated in Figure 6.1. A national master plan would control more detailed departmental planning, such as mineral and water resources, power, communication, industry and services. These plans in turn could be sub-divided into regions or river basins, and so on.

The national master plan would incorporate factors such as capital availability, population, foreign exchange, and economic growth rate. It would control or encourage outputs of the various sectors by applying shadow values. Social and environmental factors can be included although the present example refers to economic objectives only.

The objective of the individual departments would be to optimize the net economic benefit, applying the shadow values suggested by the national master plan to output. For instance the department of Water Affairs would plan to maximise the difference between value of agricultural produce, hydro-electricity and other benefits of water, and the cost of dams and conduits. The department would in turn control the transfer of water or produce between river basins by means of shadow values imposed on the river basin plans.

The whole process is ideally suited to computer solution using the principle of decomposition of linear programmes (Dantzig, 1963). Each plan

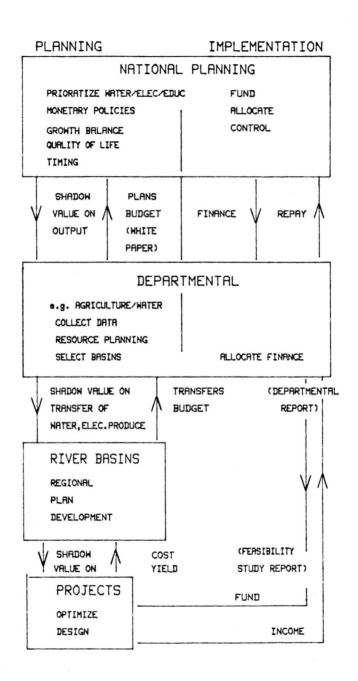

Fig. 6.1   Master Plan Flow Diagram

would be represented by the national master programme of which the departmental plans would be sub-programmes. These departmental programmes would actually also be master programmes which would have river basin or regional sub-programmes, and so on.

In fact, it would not be necessary to use linear programming to optimize each programme. Any other planning technique, such as dynamic programming, transportation programming, computer simulation or incremental analysis could be employed.

To compile the plans, much data on resources (water, mineral, manpower), economics policy, and geography will have to be stored in computer memory. A comprehensive filing index will be required for access by all departments.

The approach could be adapted to the 5 year plan concept. Capital availability would be determined after each 5 year plan, and after adjusting for private consumption and balance of payments (Kindleberger, 1965), a plan making optimum use of the available resources could be derived.

ECONOMIC POLICY

The usual criterion for selection of an optimum system is that the difference between economic benefit and cost, discounted to a common time base, is a maximum (Eckstein, 1961). This is the logical basis for planning for private enterprise, which usually wants to maximize net income, or for national planning for developed countries where market prices adequately reflect true value. However, for countries which have not yet reached full maturity, such comparison does not necessarily yield the plan which would be of most benefit to the country.

THE NATIONAL MASTER PLAN

The output of water resources development projects, namely agricultural produce, industrial and domestic water supply, hydro-electric power generation and recreational facilities, would be controlled by a national master plan. Outputs would be evaluated by the master programme in order to decide shadow values for this purpose.

The objective of the national master plan would be to optimize the economic growth rate of the country within the limits of available capital while maintaining balanced development of sectors. The national master programme would comprise a set of equations relating the production of various sectors. For instance the water requirements for irrigation would have to be limited to that available from various sources after accounting for urban and other consumption.

The master programme would comprise a weighted combination of individual sector sub-programmes. The optimum combination of sector sub-programmes would be selected by linear programming. (Linear programming is a mechanical method of selecting an optimum combination of variables from a system which could be described in terms of linear equations or constraints). The effect on the overall cost of the system is studied while successively introducing elements of each variable. That variable which would result in maximum benefit is brought into the programme and the process repeated until there is no further improvement.

By comparing alternative sector plans the national master programme computes marginal values, or shadow values, for the outputs from the sector programmes. These shadow values ensure that balance is maintained between sectors. Thus, a high shadow value would encourage production and a low one would limit production. In effect shadow values represent the value of output to the country.

The departments would then proceed to optimize their plans using the shadow values, and re-submit the plans (output and capital requirements) to the national master programme. The process of selecting sector plans and generating shadow values would be repeated until no further improvement in the national economic growth rate could be achieved.

The method of optimizing the master programme and solving the shadow values is based on the principle of decomposition of linear programmes and is identical to that for the solution of the water resource master programme. An example illustrating the technique is given in the following section.

MECHANICS OF THE WATER RESOURCE PLAN

In a similar manner to the national master programme controlling a number of sectors, each sector programme would control a number of sub-divisions. A water master programme would link a number of river-basin sub-programmes. Transfers of water or produce between basins would be controlled by shadow values imposed by the water resource master programme. (These may be in addition to the shadow values generated by the national master programme.)

An example involving the transfer of water from two river basins to an urban area will illustrate the mechanics of the decomposition technique.

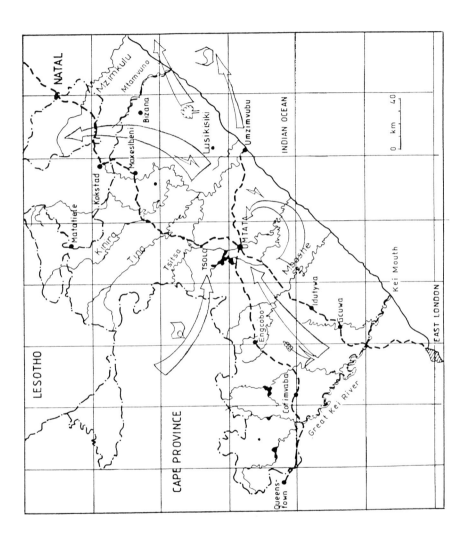

Fig. 6.2 Map of Transkei with resources allocation

The two river basins could be the Tsomo and the Mbashe, and the irrigation area of Qumanco.

Considerable detail is omitted from the example in order to highlight the method of solution. The fact that conditions will continuously change with time is ignored, and a particular time horizon is studied. The method of discounting annual cash flows and capital costs to a common time base is glossed over. Inter-basin transfer of commodities besides water, such as agricultural produce and power is likewise omitted. In fact, the system would be more like that depicted in Fig. 6.2.

At the particular time horizon chosen, the total quantity of water required from these two rivers is 636 million cubic metres per year (Mm^3/a). The objective of the master programme is to minimize total costs, at the same time considering possible irrigation and hydro-electric projects. Now suppose that some possible development plans have been proposed for each river basin. These plans could have been compiled locally or by the department of Water Affairs, and need not necessarily be optimal at this stage. All that is required by the water resource master programme is the possible quantity of water which could be diverted to the Qumanco, and the associated net cost or benefit. This data is summarized in Table 6.1 for a possible Mbashe basin project and two possible Tsomo basin projects.

TABLE 6.1 Initial Basin Programme Solutions

	Sub-programme 1		Sub-programme 2	
	Net benefit*	Diver-sion	Net benefit	Diver-sion
	Mil. Dollars/ yr	Mm^3/a	Mill. Dollars/ yr	Mm^3/a
Mbashe basin	−24.758	522		
Tsomo basin	− 6.941	116.3	+ 6.347	0

* Minus signs indicate cost, and plus signs net income. Costs in this case are converted to annual interest plus redemption payments.

Taken individually it is unlikely that any river basin plan, or sum of any plans, would yield exactly 636 Mm³/a for the Qumanco. However, a weighted combination of the sub-programmes may do so. Let v and w denote weight, or proportions, of various Mbashe or Tsomo basin plans which must be incorporated in the master programme. The total of the weights for each sub-programme should be unity.

Then $522.2v_1 + 116.3w_1 + 0w_2 = 636$

where $v_1 = 1$

$$w_1 + w_2 = 1$$

These equations constitute the master programme. Since there are three equations, there need to be three variables (v_1, w_1 and w_2 in this case) and hence the three sub-programme solutions. The objective of the master programme is to maximize the net income:

Maximize $- 24.758v_1 - 6.941w_1 + 6.347w_2$

Fortunately it will be found on solution of the master programme equations that none of the weights are negative. If they were, artificial slack variables would have to be introduced into the equations.

It may be possible to improve the objective function by introducing new sub-programmes. A simple method for finding improved solutions to the basin sub-programmes exists. Shadow values on the diversion are generated by the master programme and the basin plans are optimized using these shadow values. The shadow value is in fact the lowest capital requirement per unit of additional diversion, or

$$\$m \; \frac{6.347 - (-6.941)}{0 - 116.3} = \$114 \; 100/\text{year per Mm}^3/\text{a}$$

The shadow value, together with the national programme's shadow values, is used in optimizing the basin programmes. The benefit coefficients per unit of diversion would be the shadow value less the actual capital and pumping costs.

Plans which would maximize the net economic benefit are compiled for the individual river basins. These plans may be derived by computer program, graphically or by trial and error analysis.

It was found upon optimizing the basin programmes that an identical solution to sub-programme 1 was indicated for the Mbashe basin, whereas an alternative solution for the Tsomo basin emerged; a diversion of 83.4 Mm³/a would be possible for a total net capital expenditure of $3.144 million per year (omitting the shadow values imposed by the water resource master programme). It will be found on substituting the third Tsomo sub-programme in place of sub-programme 2, and solving for weights

v_1, w_1 and w_3, than an overall improvement in net benefit would result. It will be found by inspection that sub-programme 3 would have to replace sub-programme 2 and not sub-programme 1, in order to keep the weight w less than unity.

The water resource master programme is re-solved for shadow values, and the procedure of optimizing the basin sub-programmes repeated. However, it would be found that no further improvement to the master programme was possible, so an optimum solution is at hand. The master programme was solved to yield the following weights:

v_1 = 1.0

w_1 = 0.924

w_3 = 0.076

The weighted diversion becomes

1 × 522.2 + 0.924 × 116.3 + 0.076 × 83.4

= 636 Ml/d

of which the amount from the Tsomo basin is −

0.924 × 116.3 + 0.076 × 83.4

= 113.8 Mm^3/a

The optimum solution is depicted schematically in Figure 6.3.

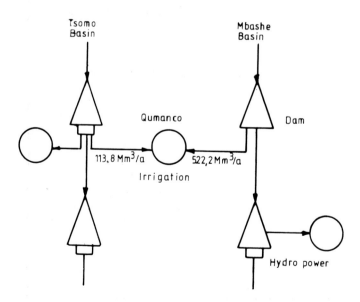

Fig. 6.3 Optimum Diversion Plan Example

127

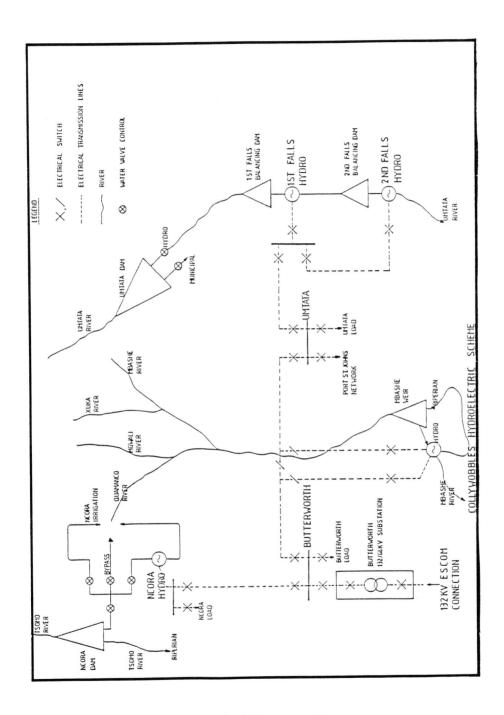

Fig. 6.4 Flow plan and electrical interconnection

The procedure for decomposition of linear programmes can be considerably streamlined, as it needs be for more complicated systems than the example. The reader should refer to Dantzig (1963), for a fuller and theoretical explanation of the technique of decomposition of linear programmes.

RIVER BASIN SUB-PROGRAMMES

For simple water storage problems it may be expedient to design the optimum plan by graphical or manual techniques. Reasonably complex projects will require the aid of system analysis techniques for formulating river basin system plans into a form suitable for computer analysis. Such techniques have been well developed, (e.g. Maass, 1962).

As developing countries hydrology differs considerably from that in developed countries with adequate data, simplistic techniques have been developed for relating reservoir yields to storage (Scheurenberg and Midgley, 1967). The extreme variability of our annual river flows means that the probability component must be adequately accounted for. The method proposed is to relate water released from reservoirs to storage capacity by storage-draft-frequency analysis. For each river basin massed flow curves or river flow records are analyzed over low-flow sequences by graphical or numerical methods to determine the storage required to meet required constant or variable drafts. The risk of failure to meet demands is determined from the frequency of the low-flow record.

The river systems is then described in terms of a set of equations or constraints. For instance, return flow may be expressed as a function of urban consumption, or if hydro-electric power generation is feasible, the relationship between hydraulic head, rate of release of water and power generated is expressed algebraically. Each variable is assigned a cost coefficient to enable an optimum solution to be selected.

The set of equations is then analyzed by linear programming or other means to achieve an optimum project design. The resulting development plans are submitted to the water resources master programme which may elect to revise shadow values on commodities transferred between river basins.

If a river basin plan is very complex, it may be subdivided into projects using the same decomposition principle.

CONCLUSIONS

The fact that the planning model can be sub-divided into independent steps is of great value. All river basins, at whatever stage of planning or development, can be included in the water resource master programme. Various projects could be analyzed and designed by different methods, such as linear programming or simulation, and still form part of the master plan. In fact, informal proposals based on experience or intuition may suffice for small projects. All that would be required is an estimated budget and the corresponding yields of water or farm produce.

The penalty to be paid for the versatility of the technique is that it is necessary to anlayze the master programme and sub-programmes successively a number of times in order to arrive at an optimum solution. However, if the programmes are retained in computer memory, repetitive solutions would be simple and rapid. In fact, with national economic growth and other new data continuously being fed in, a final solution to the whole plan may never be reached.

The advantages of planning with the aid of computers would be manifold. Large volumes of data could easily be stored and retrieved and once programmes were compiled, the possibility of making arithmetical errors would disappear. Alternative plans and their implications could easily be studied, and sensitivity analyses could be performed. Maps and tables of data could be produced by computer plotting devices and high speed printers. In fact the concept could not be realized without the use of computers.

REFERENCES

Dantzig, G.B., 1963. Linear Programming and Extensions, Princeton Univ. Press.
Eckstein, O., 1961. Water Resources Development. Harvard Univ. Press. 300p.
Kindelberger, C.P., 1965. Economic Development. McGraw Hill, N.Y. 395p.
Maas, A., Hufschmidt, M.M., Dorfman, R., Thomas, H.A., Marglin, S.A. and Fair, G.M., 1962. Design of Water Resource Systems. Macmillan, London, 620p.
Scheurenberg, R.J. and Midgley, D.C., 1967. Sequences of deficient river flow in some regions in S.A. Trans. S.A. Inst. Civil Engrs. May.
Stephenson, D., 1989. Planning model for water resources development in developing countries. Proc. Symp. IAHS. Baltimore, Publ. 180, 63-72.

CHAPTER 7

RESERVOIR SIZING

INTRODUCTION

An important element in water resources systems in many developing countries is reservoir storage. Since the hydrology of such countries is often extreme – either subject to drought or floods, reliable river yield is highly dependent on efficient reservoir sizing. The variation in monthly and yearly flows is restricted by storage, but risk of failure is a real social problem requiring careful attention in reservoir sizing and in assessing costs.

Unfortunately lack of hydrological data makes accurate reservoir sizing using simulation and stochastic means difficult. Frequently use of short term records is all that is possible.

Reservoirs are designed to store water from time of surplus for use in time of drought. Unless occasional shortfalls can be tolerated, the drawoff from or reliable yield of a reservoir must be less than the long-term mean annual river inflow. Water will be lost by spill in times of flood, and by evaporation, seepage and regulated discharge, to provide flood storage, or to supply riparian users downstream.

The best combination of reservoir size, drawoff and frequency of inability to meet the design yield, will be a matter of economic and social preference. An optimum combination can be selected using systems analysis methods but this is sophisticated and very often a preliminary estimate of reservoir capacity to meet a specified draft with a certain frequency of shortfall is required.

An analysis of existing or synthetic flow records will produce the desired reservoir capacity. The analysis can be done by computer simulation with discrete time increments in tabulation form or graphically. Direct analytical procedures for estimating storage volume are outlined later but these are simplistic and not sensitive to operational optimization. A simulation, graphically or numerically, will reveal the storage state versus time history and facilitate study of marginal variations e.g. variable draft operation, more frequent failures in supply, or changes in yield due to abnormal climatic conditions, catchment usage changes, or silting of the reservoir.

Graphical methods in particular are useful for depicting storage state histories. The disadvantage of the method is that an inflow record is necessary. However, synthetic records, probability matrix methods or analytical methods can be employed, but the data can only be as valuable as the initial data set, in particular the flow record.

Analysis based on existing records was referred to by McMahon and Mein (1978) as Critical Period analysis, because the critical dry period in the records dictates the storage volume required.

Generally methods of determining storage capacity at a particular river reach or site can be divided into four types:

(i)   Critical period techniques; based on historic flow records. Graphical or numerical analysis will reveal storage requirements as a function of draft, e.g. mass flow curves or simulation (Hufschmidt et at, 1966).

(II) Probability matrix methods: based on statistical properties of the flow variations and independent sequence.

(III) Synthetic flow sequence either from statistics or rainfall generation and simulation or analysis of the resulting flows, or use of statistical properties to derive deficits (Maass et al, 1962).

(IV) Equations (analytical method).

These methods are used for planning reservoir sizes, deciding yields of reservoirs, or operating rules or to decide future risk.

Flow data can be recorded, estimated or synthesized. By estimation is meant a deterministic analysis of available data in order to reproduce as accurately and chronologically correct as possible the flow sequences experienced. In synthetic flow patterns, on the other hand, only the statistical properties of the records are retained. The statistical properties are often sufficient to estimate storage requirements based on uniform drafts. On the other hand, frequently one wishes to consider variable draft operating rules, or the fluctuations in reservoir level are to be considered. In such case a flow record can be synthesized. A random component is input to generate a synthetic record.

DEFINITIONS

A *reservoir* is a volume of water used to draw on in times of short-fall in the river flow.

Real reservoir *storage capacity* will be limited by the topography and dam wall height and has a maximum and minimum.

Some theories assume the reservoir is *infinite* i.e. the reservior can empty but never spill. This means that all flood water is stored. Other theories assume a semi-infinite storage i.e. one which can spill but never run dry (Moran, 1959).

Active storage is that above dead storage where dead storage is inaccessible due to the level of the offtake, or is allowed for silt accumulation.

Flood storage is capacity provided, often above spillway level, for attentuating floods and cannot be relied upon for increasing the yield.

Carry-over is the amount of water stored from one time period to the next.

The *time period* is the interval between storage computations, usually monthly for water supply reservoirs where there is seasonal variation in inflow or annually for areas subject to large annual flow fluctuations.

Critical period is that in which the storage goes from full to empty without spilling.

Inflow is measured in cubic metres per month or some other suitable units. Data may be either from past flow records, estimated from rainfall and related data, or synthesized from statistics derived from existing data.

MASS FLOW METHODS

A plot of river inflow rate versus time with a draft line super-imposed will reveal when storage is required and when there is surplus water (see Fig. 7.1).

It is difficult to decide from Fig. 7.1, however, what storage capacity is required, since carried-over shortfalls must be added i.e. the total storage required is not just the area of one shaded portion — it is the sum of successive shaded portions, less inflow, provided net inflow does not exceed the reservoir capacity.

For this reason it is easier to estimate the total maximum storage requirement from a massed flow curve such as Fig. 7.2. This type of analysis is attributed to Rippl (1882).

MONTHLY FLOW   (× Mean annual runoff)

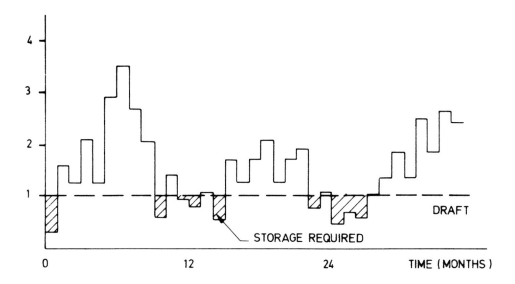

Fig. 7.1 Flow Rate versus Time Curve

The method of calculating storage is as follows: Plot cumulative inflow vertically against time from the start of the record on the horizontal axis. The slope of the curve represents the rate of inflow. A constant draft would also be represented by a positive sloping line.

If storage is just depleted at the end of a dry period, then the draft line will touch the inflow (point A). If the draft line is projected backwards, the difference between the ordinates of the draft line and the inflow line represents storage at any point in time. The maximum such difference (B) represents the storage required to meet the chosen draft for that drought period. Extending the draft line further back indicates stored volume is increasing with time here i.e. the inflow exceeds outflow. In fact the reservoir would start filling at point D and may be full by some point E. Between E and B there would be spill.

The technique may be extended to allow for evaporation or any other loss. Net evaporation loss is calculated for each time period (usually monthly) by multiplying net evaporation rate by reservoir water surface area existing at the time. This is added to draft to make total loss each month, and the storage required now becomes CG.

The same approach may be used in tabular or computer solutions.

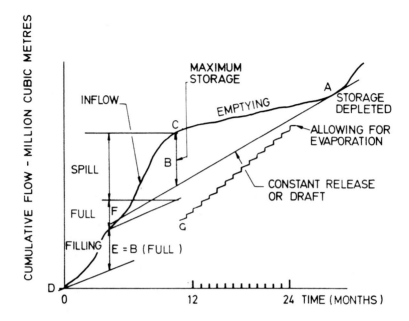

Fig 7.2 Massed Flow Curve for Storage Analysis

SIMULATION OF RESERVOIR OPERATION

In order to investigate the adequacy of a water resource system design, operation of the system can be reproduced, or simulated, numerically. The analysis can be done by hand, studying only a short critical length of stream-flow record, or by digital or analogue computer capable of handling a large amount of data with relative ease.

A computer programme may be devised for performing reservoir simulations by repetitive application of the hydrological equation.

$$S_{n+1} = S_n + I_n - U_n - E_n - [F_n] \qquad (7.1)$$

where S_n is storage at the beginning of a month n

I_n is inflow for month n

U_n is release for month n

E_n is evaporation, which is a function of S_n

F_n = Flood overflow. [ ] implies this term is omitted if not positive

It is convenient to approximate the reservoir surface area by a mathematical function of stored volume, and to express evaporation loss as the product of mean net evaporation per unit area for the month and reservoir surface area. Seepage and channel losses may also be accounted for in the analysis.

It is usual to work with monthly flow records for draft studies, and with daily or even hourly readings for flood studies. Inflows I for a series of months may be obtained from records and fed into the computer.

Draft U is specified as a constant value or as a mathematical function of storage state.

The simulation procedure is commenced by specifying an initial storage and working through the available record month by month. Water demand level is held constant for a particular time horizon and historic flows routed through the system. The process may be repeated for demand levels associated with different time horizons.

It is specified in the programme that if storage state reaches the capacity of the reservoir, spill will occur. Storage state at each month can be printed or plotted by the computer. If at any stage the storage is emptied, the design is deemed inadequate and storage capacity should be increased.

An optimum system design can be derived by analysing a number of combinations of reservoir capacity and draft. Net benefits are computed for each case, and that trial which indicates optimum benefit is selected. Variable draft operating rules can readily be accommodated in the programme. A complete picture of storage states and drafts results from the analysis.

If a system comprising a number of reservoirs is to be simulated, then a large number of possible designs may exist, and many computations may be needed if adequate coverage is to be achieved. It is in such instances that the advantage of high speed computers becomes evident.

When simulation methods are used, upper and lower bounds to storage can readily be accommodated. For example, draft can be stopped when storage is empty. This is not the case if the reservoir is assumed semi-infinite. Some analytical methods described later use the assumption of semi-infinite or even infinite storage. In the case of infinite storage all flood water is retained and the calculations will over-estimate yield. If semi-infinite storage is assumed then the probability or frequency of failure is overestimated.

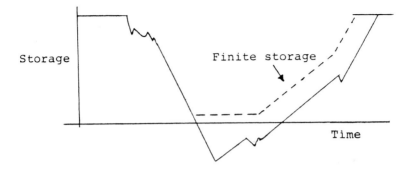

Fig. 7.3 Storage – simulation assuming semi–infinite storage

STORAGE–DRAFT–FREQUENCY–ANALYSIS

The storage required to secure a desired yield from a river can be determined by iterative solutions of the hydrological equation. These may be performed either graphically or numerically. The graphical method is often referred to as the mass diagram technique, usually attributed to Rippl (1883), and the latter as the water budget technique. Both are described in standard hydrological textbooks e.g. Viessman et al (1977).

If the record of riverflow at a proposed reservoir site is long enough to be reasonably representative of the full range of flow conditions likely to be experienced, the storage–yield relationship resulting from a straightforward massed flow analysis of the record might be considered acceptable as a basis for reservoir design but would be subject to the proviso that drought sequences of the future would not be more severe than any in the record on which the analysis was based. In this method, the frequency with which a given storage would fail to meet a given yield would not be revealed. Methods have therefore been developed whereby the recurrence interval of deficient flow during the critical period associated with a desired yield from storage can be determined and in this way the failure frequency aspect of storage design can be introduced.

Storage–Draft Calculations

The degree to which storage requirements to meet a given duty are controlled by critical deficient inflow sequences is best explained with reference to the storage equation.

Let:

C be the full supply capacity of storage provision

S_t the storage state at time t measured from an arbitrary time origin O

$\sum_o^t O$ the cumulative outflows or withdrawals (including losses) from storage to time t

$\sum_o^t I$ the cumulative inflows to storage to time t

$\sum_o^t W$ the cumulative spillages from storage to time t

$\sum_o^t D$ the cumulative deficits, i.e. the total volume by which the system short-supplied O_t, to time t.

The storage equation can then be written:

$$S_t = \sum_o^t I - \sum_o^t O - \left[\sum_o^t W - \sum_o^t D \right] \qquad (7.2)$$

It follows that storage of capacity C will meet a given total drawoff (including losses) without deficit or spill throughout all time intervals $(t_2 - t_1)$ during which:

$$\sum_{t_1}^{t_2} I \geq \sum_{t_1}^{t_2} O - C \qquad (7.3)$$

For given values of the right-hand side of the above equation there are thus critical values of the left-hand side i.e. critical inflows associated with any period $(t_2 - t_1)$ during which neither shortfall nor spill will occur. If the frequencies of occurrence of various values of inflow associated with prescribed lengths of time interval $(t_2 - t_1)$ can be ascertained, it follows that the frequency with which storage of capacity C can meet a specified drawoff rate can be established by testing of the

138

inequality:

$$\sum_{t_1}^{t_2} I \geq \sum_{t_1}^{t_2} O - C \qquad\qquad (7.4)$$

Testing is necessary because total withdrawals, O, include drafts for use as well as evaporation losses and separation of use from loss is complicated by the fact that evaporation is dependent upon both storage state and time of year.

The basis of the method is to treat deficient flows as extreme values, to abstract low flows of various durations from the record and analyse them by statistical techniques.

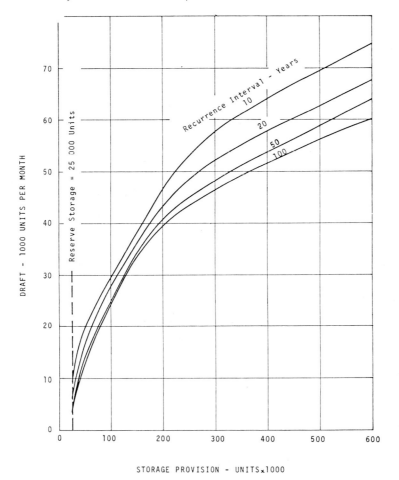

Fig. 7.4 Storage – Draft – Frequency Curves

Selection of Low-Flow Sequences

Whether considering rainfall, runoff, storage requirement or other such feature, statistical analysis begins with a ranking, or arrangement in order of magnitude, of extremes selected from the recorded data.

Independent minimum flow sequences are chosen from an array of cumulative flows for each duration. The minimum flows may be arranged accordingly to one of several types of series, of which two are:

(i)   Discrete time series – one event per time unit (usually one year) is chosen

(ii) Partial duration series – as far as possible to ensure independence, overlapping sequences are eliminated after selection of a particular sequence (see e.g. Midgley, 1967).

The discrete time series is unsuitable for drought-flow analysis in which durations of up to 8 years are usually considered. Much valuable information about low-flows for short durations would be lost if the time were chosen sufficiently large to suit the long durations whereas, on the other hand, if a different time unit were to be used for each duration, the results would be inconsistent. The main objection to the discrete time series, however, is that a particular severe drought sequence might fall partly within one time interval and partly in the next, in which case it would be missed when the critical low-flows are selected.

In the light of the foregoing the partial duration series offers the better solution when selecting sequences of long duration while the discrete time series is best employed when selecting sequences of short duration. Thus for sequences from one up to, say, nine months' duration the severest sequence from each low-flow season (i.e. one per year) is selected. In the analysis of drought sequences lasting twelve months or longer, the partial series is adopted, and overlapping is eliminated by the screening procedure described below. Carried out manually, however, the procedure is extremely laborious and is best performed by computer.

Frequency Analysis

The next step in the analysis is to assign probabilities to the ranked values. The method most commonly employed is to assign a recurrence interval T in years according to the Weibull equation:

$$T = \frac{n+1}{m} \qquad\qquad (7.5)$$

in which n is the total number of years of record under analysis

m is the rank of the sequence in the ascending array.

The inter-relationship of recurrence interval, duration and inflow is best determined by plotting the results on a specially-ruled probability paper suggested by Gumbel (1958). The recurrence interval scale of Gumbel paper can be calculated from the formula:

$$y = -\log_e \left[-\log_e \left(1 - \frac{1}{T} \right) \right] \qquad\qquad (7.6)$$

in which y is Gumbel's *reduced variate*, the values of which can be plotted to a linear scale.

Extension to Ungauged Areas

A storage draft curve is specific to the reservoir for which it was constructed. Even if the storage and draft variables could be rendered dimensionless (e.g. by expressing them as percentages of mean riverflow), it is unlikely that the resulting diagram would be representative of other basins within the same hydrological region because storage/evaporation functions would probably differ markedly from reservoir to reservoir. If regional generalization is to be attempted it will clearly be necessary to handle the influence of evaporation separately.

Cumulative inflows to storage during drought sequences having 5-, 10-, 20-, 50 and 100-year return periods, each commencing January through December, may be expressed in terms of mean annual runoff (as per cent MAR) and analysed to yield the storage (as per cent MAR) required at each month of the year to maintain a given gross draft (as per cent MAR) at various levels of assurance, ignoring evaporation losses.

The contention is that, inasmuch as it has been found that the dimensionless cumulative inflow curves (critical mass curves) can be generalized within hydrologically similar regions, diagrams can likewise be regionally grouped and it remains then to develop a means for making the allowance for evaporation so that gross drafts may be converted to net drafts.

RESERVOIRS IN SERIES AND PARALLEL

For an isolated reservoir on a river the draft as a fraction of the mean annual river flow, i.e., q_1/F_1, is read from the draft–storage curve corresponding to the point with an abscissa equal to d_1/F_1, in which S_1 = reservoir capacity and F_1 = mean annual river flow. Where two or more dams are built in series, (d_2 being the downstream reservoir, and F_2 the mean inflow between reservoirs d_1 and d_2) the following holds:

First calculate Q_1/F_1 for S_1/F_1, then;

1. If $S_2/F_2 \leq S_1/F_1$ then q_2/F_2 is indicated by the point on the draft–storage curve with an abscissa equal to S_2/F_2. Thus, the total draft available at the downstream site is q_2 plus whatever draft from the upstream reservoir is not used consumptively.

2. If $S_2/F_2 \geq S_1/F_1$ then $(q_2 + q_1)/(F_2 + F_1)$ is indicated by the point on the draft–storage curve with an abscissa equal to $(S_2 + S_1)/(F_2 + F_1)$.

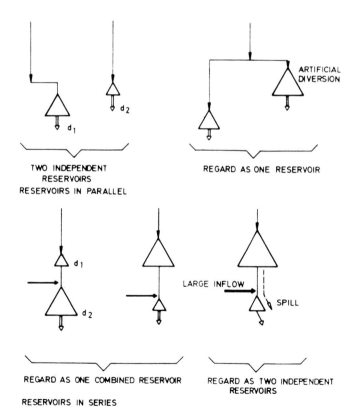

ARTIFICIAL DIVERSION

TWO INDEPENDENT RESERVOIRS

RESERVOIRS IN PARALLEL

REGARD AS ONE RESERVOIR

d_1

d_2

LARGE INFLOW

SPILL

REGARD AS ONE COMBINED RESERVOIR

REGARD AS TWO INDEPENDENT RESERVOIRS

RESERVOIRS IN SERIES

Fig. 7.5 Reservoirs in Parallel and Series

The criterion strictly holds only if inflow between the two reservoirs is in constant proportion to inflow upstream of reservoir 1. Also, the evaporation versus storage characteristics for the two reservoirs should be similar. With tributaries rising in neighbouring catchment areas these conditions are usually satisfied for all practical cases. If not, different draft-storage curves could be prepared for each site (Stephenson, 1970).

It is necessary to guess the S/F ratios for various reservoirs prior to the design, in order to know which relationship is applicable. If the guess proved too far out, the relationship may have to be altered according to a trial and error process. Fortunately for most practical cases results are relatively insensitive to the choice made.

STOCHASTIC FLOW

Analytical and Synthetic Flow Generation

The statistical properties of flow records can be used to generate new records. As long as it is recalled that the resulting sequence cannot give more information than the original data used to derive the statistical record, the technique can be of great use. Both analytical (equation form) results and new longer flow records can be prepared from limited stream flow measurements or records. Analytical results can be used for quick estimates of storage or yield, and long synthetic records can be used to simulate reservoir operation for alternative operating conditions. Both can be used to estimate risk of failure.

Statistical Analysis Definitions

Arithmetic mean
$$\bar{x} = \frac{\sum_{1}^{n} x_i}{n} \tag{7.8}$$

Median: middle value of the variate which divides the flow frequency distribution into 2 equal portions.

Standard deviation $s = \left[\frac{1}{n-1} \sum_{i}^{n} (x_i - \bar{x})^2 \right]^{1/2}$ \hfill (7.9)

$$= \left[\frac{1}{n-1} \left(\sum_{i}^{n} x_i^2 - n\bar{x}^2 \right) \right]^{1/2} \tag{7.10}$$

Variance $\quad\quad\quad = s^2$ \hfill (7.11)

Coefficient of variance

$$\text{COV} = s/\bar{x} \qquad (7.12)$$

Skewness
$$C_s = \frac{a}{s^3} \qquad (7.13)$$

$$a = \frac{n}{(n-1)(n-2)} \sum_i^n (x_i - \bar{x})^3 \qquad (7.14)$$

$$= \frac{n}{(n-1)(n-2)} \left[\Sigma x^3 - 3x \Sigma x^2 + 2n\bar{x}^3 \right] \qquad (7.15)$$

Serial Correlation
$$r = \frac{\Sigma(xy) - \frac{\Sigma x \Sigma y}{n}}{\sqrt{\Sigma x^2 - \frac{(\Sigma x)^2}{n}} \sqrt{\Sigma y^2 - \frac{(\Sigma y)^2}{n}}} \qquad (7.16)$$

$$= \frac{\displaystyle\sum_{i=2}^{n} Q_i Q_{i-1} - (n-1)\bar{Q}^2}{(n-1)s^2} \qquad (7.17)$$

SYNTHETIC FLOW RECORDS

Methods for storage calculation are similar to previous except that input is not an actual record, it is synthesized (Brittan, 1961).

Streamflow can be regarded as consisting of four components as in Figure 7.6 Paling and Stephenson (1988) further propose cyclic variations.

The Chow model is:

$x = \bar{x} + st$

The Markov model is:

$$x_{i+1} = \bar{x} + r_1(x_i - \bar{x}) + t_i s (1 - r_1^2)^{1/2} \qquad (7.18)$$

where r_1 = annual lag-one serial correlation coefficient

t_i = normal random variate with zero mean and unit variance

Note we sometimes get negative flow this way, which could be set equal to zero. To account for skewness one could use Gamma distribution (this needs to be done numerically) or assume log Q is normal – then this needs to be solved iteratively. It also eliminates negative flows.

The method is easiest to apply to annual flows. Annual flows exhibit little serial correlation, but monthly variation needs separate treatment since seasonal variation must be accounted for.

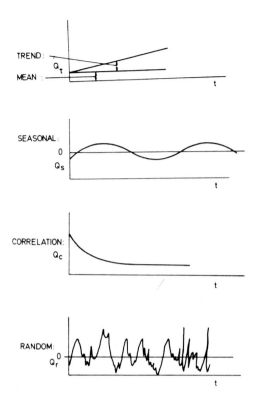

Fig. 7.6 Flow Components

ESTIMATING STORAGE WITH EXTREME VALUE DISTRIBUTION AND SERIAL CORRELATION

Methods of estimating storage may use existing flow records directly or statistics derived from flow records. Into the former category falls the massed flow curve method of Rippl (1883). On the other hand Hazen, (1914) and Sudler (1927) introduced probability concepts with the method. Critical periods are identified by graphical or numerical simulation methods. The probability of failure can be estimated by ranking the droughts as indicated by Stall (1962). Recurrence intervals may be calculated using equations such as Gringorten's (1963). Systems analysis methods, (Klemês, 1979) may be employed for appraisal of the likelihood of failure but in general massed flow techniques are only suitable for selecting the safe yield during the repeated inflow sequence. The method is most appropriate for the existing records. Records can be extended and generated synthetically (Fiering, 1967), if necessary, although the resulting record is only as reliable as the initial data. Storage is estimated by trial and error for any selected draft. A separate analysis is required for each site and draft, and these analyses are time-consuming. They require considerable data in the way of flow records at the site in question.

Direct estimation of reservoir storage capacity is possible using probability matrix methods (Moran, 1959) or analytical methods. Alexander (1962) presented reservoir capacity charts assuming river flows obeyed a Gamma distribution. McMahon and Mein (1978) point out that the theory can be employed as a first estimate and the design should be checked and refined using simulation methods. This is because the Gamma distribution fit is not always accurate and serial correlation is not easily accounted for. In all cases uniform draft is assumed.

Reservoir capacity functions such as those of Alexander facilitate design since a complete record is not required and calculations are quick and simple. Statistical data such as means and coefficient of variation (and serial correlation in the present approach) are sufficient to yield reservoir capacity for any desired level of draft and probability of failure. The proposed method follows these lines, enabling reservoir capacity to be selected with a minimum of data and effort. The accuracy of the answer approaches that of massed flow methods and an additional variable, namely, probability of failure, is introduced. The parameters required, flow mean, variance and serial correlation are readily obtained, and the resulting storage draft curves can be applied either on a regional basis or for a specific site.

Annual Flow Distribution

Reservoirs are frequently required to store water from one year to the next on account of the extreme flow variations between years. An analysis of mean flow on an annual basis is therefore required for major reservoir design. If the flow-frequency distribution can be approximated by a mathematical distribution, this facilitates analytical solutions for the storage function. Alexander (1962) found a Gamma distribution approximated the annual flow distributions in Australian rivers. Fiering (1967) suggested a log-normal distribution after realising that a normal distribution resulted in a large number of negative flows in synthetic records. The Extreme Value Type III (Weibull) distribution for minima as explained by Haan (1977) also avoids negative flows, but none of these distributions could be fitted accurately to annual river flows. A reasonable fit was obtained assuming an Extreme Value Type I distribution as proposed by Gumbel (1954) for peak flows. For low flows it is necessary to replace the probability of exceedance with the probability of being less than. It is also necessary to replace the arithmetic mean by the 1.75-year value which is the theoretical mean and estimate the coefficient of variation using only the annual low flows (i.e. less than the adjusted mean which is the 1.75-year value). In fact an eye fit on a Gumbel plot is the most satisfactory method.

The extreme value distribution is an approximation to the tail-end of an exponential distribution and also fits the tail of a normal distribution. The distribution can be described in terms of the mean and standard deviation. The resulting equation for flows enables one to derive analytically a relationship between flow shortfalls and reservoir capacity.

The extreme value distribution is intended to fit only the tail of a complete distribution. If the two tails differ in shape, then the appropriate one only (low flows in this case) is the one which should be analysed to obtain the coefficient of variation. The distribution also theoretically passes through the mean at the 1.75-year value. If annual high flows are used in estimating the mean, then owing to skewness, the resulting mean may not be the best fit for the low-flow distribution. The equation for the distribution may also differ from the true distribution near the mean. This is the reason for selecting the value of the effective 1.75-year value rather than the true mean.

Figure 7.7 indicates the low values of annual discharges for Vaal river at Vaaldam, taken from a 43-year record. The equation $T = (n+1)/m$ was used to calculate plotting positions, where T is the recurrence interval (inverse of probability of being less than), n is the number of years of record and m is the rank of the discharge starting from 1 for the lowest value.

A Type I and Type III extreme value distribution line were also plotted using the arithmetic means and coefficients of variation and the appropriate equation e.g. (1). The preferred line is an eye fit through $Q_{1\cdot75} = 1600 \times 10^6 m^3$ and $Q_{50} = 400 \times 10^6 m^3$. This approximates the actual record and its trend (graph slope) for recurrence intervals of interest (20 to 50 years in this situation proved an economic return period of failure).

The Extreme Value Type I distribution indicates negative discharge for extreme recurrence intervals. For a coefficient of variation of less than 0.5 (normally the case) the recurrence interval of negative flows is greater than 1500 years, which is too remote to be a problem for the simplistic design method proposed here.

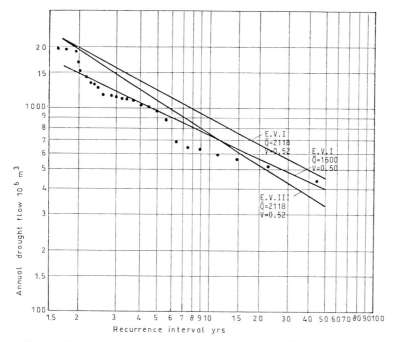

Fig. 7.7 Extreme value plot of annual river flows

EXTREME VALUE DISTRIBUTION

The extreme Value Type I distribution, henceforth here referred to as the Extreme Value distribution, may be written as follows for low flows:

$$Q_T = Q - sK(T) \tag{7.19}$$

$$\text{where } K(T) = (\sqrt{6}/\pi)\ (\gamma + \ln \ln T) \tag{7.20}$$

is referred to as the frequency factor $K(T)$.

γ is Eulers constant, 0.57721. Q is the mean annual flow, and s is the standard deviation of annual flows, $[\ \Sigma\ (Q-\bar{Q})^2/(n-1)]^{1/2}$ T is the recurrence interval of the drought flow being equal to or less than Q_T. A correction factor should be applied to K for small samples (Viesmann, et al, 1977).

If it is assumed that a reservoir is full at the beginning of a drought, then the storage required to meet a deficit for one year

$$V = D - Q_T \tag{7.21}$$

$$= D - Q + sK(T) \tag{7.22}$$

where D is the demand or draft during the period.

It may occur that the critical drought i.e. one which requires most storage, for any particular recurrence interval T, is longer than 1 year.

The average flow over N years may be estimated from the theory of samples. The expected mean of a sample is $Q_N = Q$ and the standard deviation of the mean is $S_N = s/\sqrt{N}$ where N is the number of years in the sample.

If it can be assumed that the Extreme Value distribution approximates the tail of a normal distribution, then the extreme (low value) means of N-year samples will also have Extreme Value distribution. This follows from the Central Limit theorem which states that the full sample of means will be normally distributed. Then the T-year sample is by analogy with (7.18)

$$(Q_N)_T = Q_N - S_N K(T) \tag{7.23}$$
$$= Q - (s/\sqrt{N})K(T) \tag{7.24}$$

Hence expected total flow over N years of sample is

$$N(Q_N)_T = N \quad Q - (s/\sqrt{N})K(T) \tag{7.25}$$

The storage required to meet the shortfall over N years is

$$(V_N)_T = N \quad D-Q+(s/\sqrt{N})K(T) \tag{7.26}$$

It is necessary to compare the storage for alternative drought periods (N = 1.2...) before selecting the maximum or to differentiate and find the most.

Serial Correlation

Although the 1-year drought for any T is the most severe, the cumulative deficiency may be more severe for longer periods. This comparison may be performed graphically for any value of T by plotting $(D-Q)/s$ against V/s for different N values. It will usually be found that the 1-year drought is critical for serially uncorrelated flows. It will be found in practice that this is often not the case i.e. a drought is likely to persist and may require storage for a number of years. Annual flows can be correlated mathematically with the serial correlation coefficient. The lag-one serial correlation is

$$r = \frac{\sum\limits_{2}^{n} (Q_i Q_{i-1}) - (n-1)Q^2}{(n-1)s^2} \tag{7.27}$$

The following Markov model was proposed by Brittan (1961) for lag-one serially correlated flows:

$$Q_i = Q + r(Q_{i-1}-Q) + (1-r^2)^{1/2} st \tag{7.28}$$

where t is a normal random variate with a mean of zero and standard deviation of unity. It is reasonable to replace the product st corresponding to T by $Q_T - Q$. Then substitution of $Q_T - Q = -sK(T)$ into (7.28) yields

$$(Q_i)T = Q + r(Q_{i-1,T} - Q) - (1-r^2)^{1/2} \; sK(T) \tag{7.29}$$

If it can be assumed that the same K applies to the T-year sequence as occurs to the individual flows (this is reasonable as K represents the level of drought) then the mean flow over the sequence is

$$(Q_N)_T = Q - r \; s \; K(T) - (1-r^2)^{1/2} \; (s/N^{1/2})K(T) \tag{7.30}$$
$$= Q - \{r + (1-r^2)^{1/2} / N^{1/2}\} sK(T) \tag{7.31}$$

The total flow over a critical sequence of N years is
$$N(Q_N)_T = N[Q - \{r + (1-r^2)^{1/2} / N^{1/2}\} \; sK(T)] \tag{7.32}$$
The storage required to meet a steady draft rate of D per annum over N years against the T-year recurrence interval drought is therefore
$$(V_N)_T = N[D - Q + r + (1-r^2)^{1/2} / N^{1/2} \; sK(T)] \tag{7.33}$$
rearranging,
$$\frac{D-Q}{s} = \frac{V}{Ns} - \{r + (1-r^2)^{1/2} / N^{1/2}\} \; K(T) \tag{7.34}$$

which is more suitable for a dimensionless graphical representation of (D - Q)/s versus V/s. Again it is necessary to investigate alternative drought durations (N = 1,2, ...) before selecting the worst drought sequence, i.e. that requiring the greatest storage volume to meet a draft. The resulting storage-draft relationship is a convex upward curve illustrating diminishing returns for greater storage. It is applicable to all reservoirs within drought regions with the same r. Different curves could be plotted for different recurrence intervals T.

Figure 7.8 illustrates a storage-draft graph for a lag-one serial correlation r = 0.15. This is applicable to the Vaal river at Vaaldam in South Africa. The mean annual river flow is 2100 $\times$ 10^6 m^2/year, and the coefficient of variation is 0.7. These coefficients were calculated from annual flow records using a pocket calculator. From Figure 7.7, however, a best-fit line to low flows indicates a value for $\bar{Q}$ of 1600 $\times$ 10^6 m^3 which coincides with the 1.75 year value, and a corresponding CV of 0.5, are most appropriate.

Figure 7.8 is plotted from equation 7.34 with r = 0.15 and T = 100 years and different trial N values. Now storage V required to meet a draft D safe against the 100 year drought may be read from the figure (or vice-versa). Thus for a storage capacity of $2000 \times 10^6 \, m^3$, then $V/s = 2000 \times 10^6/(0.5 \times 1600 \times 10^6) = 2.5$. From the storage-draft curve $(D-\bar{Q})/s = -0.50$. Hence $D = -0.50 \times 0.5 \times 1600 \times 10^6 + 1600 \times 10^6 = 1200 \times 10^6 \, m^3/a$, and the critical drought duration is 10 years.

Evaporation and seepage are neglected here. They could only be accounted for by including them in part of the draft.

RESERVOIR SIMULATION WITH SYNTHETIC DISTRIBUTION

The Gumbel extreme value equation yields an explicit equation for annual flow in terms of the recurrence interval. Thus if the annual flow mean, $\bar{Q}$, standard deviation, s, and lag-one serial correlation coefficient, r, are known, then if a random number p between 0 and 1 is generated, a random annual flow sequence may be computed from the equation

$$Q_i = \bar{Q} + r(Q_{i-1} - \bar{Q}) - (1-r^2)^{1/2} K(1/p) \qquad (7.35)$$

The mean, $\bar{Q}$ may be employed as Q_o to seed the flow sequence, and a seed will also be required for the random number generator. Then a large sequence of flows should be generated to eliminate the effects of the seeds.

A normal random number may be generated on a computer or even on a pocket calculator using for example the following algorithm for mean μ and standard deviation σ.

$$RN_i = \mu + \sigma \, (-2 \ln U_i)^{1/2} \cos(360°U_i) \qquad (7.36)$$

where $U_i = FP(9821U_{i-1} + 0.211327)$, a uniform random number between 0 and 1.

If the Q's are annual flows, monthly flow could be generated using distribution functions such as

$$Q_m = (Q_i/12)f(m) \qquad (7.37)$$

where $f(m) = 1 + M \sin (2\pi m/12)$

with month $m = 1$ being at the commencement of the wet season and M being the relative amplitude of the monthly variation
$(0 \leq M \leq 1)$ $\qquad (7.38)$

Such a synthesized flow sequence cannot be employed to reproduce accurately floods as well as droughts, as the Gumbel theory applies only to one or the other extreme. Nevertheless it is a valuable tool for reservoir sizing or operational policy decision-making. The simulation may be performed on a pocket calculator provided the reservoir operating rule is not too sophisticated. Thus draft could be some function of storage state.

An optimized design or operating rule will require cost functions. The computations will increase but trial and error methods are very feasible.

Note that evaporation, spill or seepage are not accounted for except if included in the draft D by mean values.

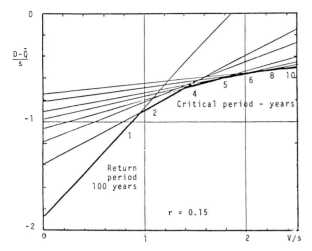

Fig. 7.8 Extreme – value Draft – Storage curve for r = 0.15, T = 100 year

PROBABILITY MATRIX METHODS

A Simple Mutually Exclusive Model (McMahon and Mein, 1978)

Provided the statistics of a flow record are available, the expected operation of a reservoir can be predicted using probability theory. Choose the inflows, draft, and storage capacity as integer multiples of some arbitrary volume unit. The following example demonstrates the technique.

Assume reservoir capacity is 2 units and a constant draft of 1 unit per time period. Inflows are discrete and independent and distributed as in Fig. 7.9. Note that the sum of the probabilities equals unity.

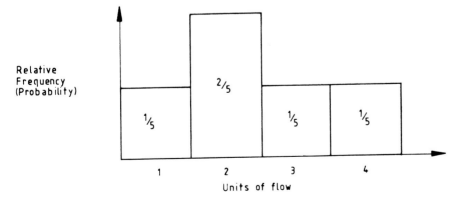

Fig. 7.9 Distribution of reservoir inflows.

For the mutually exclusive model we have:

Insufficient water $Z_{t+1} = 0$ if $Z_t + X_t \leq M$ (7.40)

Normal $Z_{t+1} = (Z_t + X_t) - M$ if $M < Z_t + X_t < K$ (7.41)

Overflow $Z_{t+1} = K - M$ if $K \leq Z_t + X_t$ (7.42)

(Assumes first inflow then release)

where Z_t = stored water at the beginning of the tth period,

Z_{t+1} = stored water at the end of the tth period or at the beginning of the $(t+1)^{th}$ period,

K = capacity of reservoir,

X_t = inflow during t^{th} period,

M = constant volume released at the end of the unit

The first step is to set up the "transition matrix" of the storage contents. A transition matrix shows the probability of the storage finishing in any particular state at the end of a time period for each possible initial state at the beginning of that period. The transition matrix for the above example is a (2 × 2) matrix representing an empty condition and a half full condition as follows:

		Initial State Z_t		
		Empty 0	1	Full 2
Finishing State Z_{t+1}	Empty 0	$\frac{1}{5} \cdot \frac{2}{5}$	$\frac{1}{5}$	0
	1	$\frac{1}{5} \cdot \frac{1}{5}$	$\frac{2}{5} \cdot \frac{1}{5} \cdot \frac{1}{5}$	1
	Full 2			0
	Σ =	1	1	1

(7.43)

(always check)

Each element of the transition matrix is found by applying Eqs. 7.40 to 7.42 to determine the inflows (and hence probability) of the storage beginning and ending in the state corresponding to that element. In the computations the boundary conditions (empty and full) must be considered and, for the mutually exclusive model, the inflows must be considered separately and prior to the outflows.

Consider the element in Eq. 7.43 which represents a reservoir starting empty and finishing empty. This can happen if there are no inflows for the period (probability 1/5) or if there is one unit of inflow (probability 2/5). In the latter case the release of one unit reduces the reservoir contents back to zero. Hence, if the reservoir starts empty there is a probability of 0.6 that it will still be empty at the end of the time period.

Consider now the element (1.0) which represents a reservoir starting empty and finishing half full. If there are two units of inflow (probability 1/5) followed by one unit of release, the reservoir will finish half full. If there are three units of inflow (probability also 1/5), the reservoir will spill because its capacity is only 2 units, then after 1 unit of release, it will again finish half full. Thus the probability of going from empty to half full is 2/5.

Note that the reservoir can never finish (and hence start) in the full condition because of the mutually exclusive assumption about inflows and outflows. Note also that the reservoir must finish in some condition; thus the sum of the probabilities in any column must be unity.

Let us now assume that the time unit is equal to one year and that the reservoir of capacity 2 units is empty at the beginning of the year one, that is, the initial probability distribution of storage content is:

$$
\begin{array}{cc}
& 0 \\
\text{Storage} & 1 \\
\text{State} & 2
\end{array}
\quad
\begin{bmatrix} 1 \\ 0 \\ - \end{bmatrix}
\qquad\qquad (7.44)
$$

$$\Sigma = 1$$

Since the transition matrix expresses the conditional probability of final storage contents given the various values of initial contents, the probability distribution of final contents can be found by the matrix product of the transition matrix and the probability distribution of initial contents. Therefore, at the end of year one (or at the beginning of the year two) the probability of storage content will be:

$$
\begin{bmatrix} 0.6 & 0.2 \\ 0.4 & 0.8 \end{bmatrix}
\begin{bmatrix} 1 \\ 0 \end{bmatrix}
=
\begin{bmatrix} 0.6 \times 1 + 0.2 \times 0 \\ 0.4 \times 1 + 0.8 \times 0 \end{bmatrix}
=
\begin{bmatrix} 0.6 \\ 0.4 \end{bmatrix}
\qquad (7.45)
$$

transition	state	state of storage	Σ 1.0
matrix	of	at end of year one	
	storage		
	at		
	beginning		
	of year one		
	(given)		

The quantitative process in Eq. 7.45 may be described as follows. The transition matrix shows the probability of the reservoir finishing in a specific state, given an initial state. If the initial state is known in terms of probability, then the joint probability will indicate the likelihood of the storage ending in a specific state. In Eq. 7.45 the transition matrix shows the probability of going from state 0 – state 0 as 0.6, and the probability of being in state 0 at the beginning of year one is 1, thus the probability of ending in state 0 is 0.6 × 1 = 0.6. But it is also possible to arrive at state 0 from state 1 which from the transition matrix has a probability of 0.2. The likelihood of being in state 1 at the beginning of the year one is 0, thus the probability of ending in state 0 but beginning in state 1 is 0.2 × 0 = 0. Hence the combined probability of ending in state 0 at the end of the first year is 0.6 + 0 = 0.6. A similar argument holds for state 1.

The process can now be repeated, using the state vector as the new starting condition. Therefore, at the end of the second year, the probability of storage content will be:

$$\begin{bmatrix} 0.6 & 0.2 \\ 0.4 & 0.8 \end{bmatrix} \begin{bmatrix} 0.6 \\ 0.4 \end{bmatrix} = \begin{bmatrix} 0.6 \times 0.6 + 0.2 \times 0.4 \\ 0.4 \times 0.6 + 0.8 \times 0.4 \end{bmatrix} = \begin{bmatrix} 0.44 \\ 0.56 \end{bmatrix} \qquad \Sigma = 1.00 \qquad (7.46)$$

transition matrix   state of storage at end of year one or beginning of year two   state of storage at end of year two

$$\begin{bmatrix} 0.6 & 0.2 \\ 0.4 & 0.8 \end{bmatrix} \begin{bmatrix} 0.44 \\ 0.56 \end{bmatrix} = \begin{bmatrix} 0.6 \times 0.44 + 0.2 \times 0.56 \\ 0.4 \times 0.44 + 0.8 \times 0.56 \end{bmatrix} = \begin{bmatrix} 0.38 \\ 0.62 \end{bmatrix}$$

$$\Sigma = 1.00 \qquad (7.47)$$

At the end of the fourth year, the probability of storage content will be:

$$\begin{bmatrix} 0.6 & 0.2 \\ 0.4 & 0.8 \end{bmatrix} \begin{bmatrix} 0.38 \\ 0.62 \end{bmatrix} = \begin{bmatrix} 0.6 \times 0.38 + 0.2 \times 0.62 \\ 0.4 \times 0.38 + 0.8 \times 0.62 \end{bmatrix} = \begin{bmatrix} 0.35 \\ 0.65 \end{bmatrix} \quad (7.48)$$

$$\Sigma = 1.00$$

At the end of the eighth year the probability of the storage content will be

$$\begin{bmatrix} 0.33 \\ 0.67 \end{bmatrix} \qquad (7.49)$$

At the end of the ninth period it will be:

$$\begin{bmatrix} 0.33 \\ 0.67 \end{bmatrix} \tag{7.50}$$

It will be noticed that as successive years are considered, the probability vector of storage content becomes less affected by the initial starting conditions (in this example, the reservoir was assumed empty) and approaches a constant or steady state situation, which is independent of the initial conditions. From the steady state vector (Eq. 7.13) it is seen that there is a 1/3 chance that the reservoir will be empty at the end of any year.

QUEUING THEORY   (Langbein, 1958)

For the steady state case a direct solution of the probability matrix is feasible. Queuing theory enables fluctuations in reservoir storage level to be correlated with statistical streamflow variations. The theory is so named because inflow to a reservoir is analogous to the random arrival of people to join a queue which is being served according to a prescribed rule. The arrivals (inflows) will conform to some statistical function of time, and the rate of serving (release of water from storage) can likewise be expressed as some mathematical function. The length of queue (volume of water in storage) must therefore be some function of inflow.

A simple numerical method of determining the likelihood of operating at various levels of storage will serve to explain the concepts.

Operation of a reservoir according to a variable draft pattern will be studied. Figure 7.9 indicates a specified operating rule, viz. normal draft is 60% of mean annual riverflow (MAR), and, once storage falls beflow 40% MAR, draft is dropped to 40% MAR. If storage should drop to zero, it will be necessary to lower the draft to 20% MAR, i.e. minimum river flow. A decision as to the rate of release is to be taken only once a year, say on the first day of October, and the draft is held constant until that day the following year, when a new decision may be made. In the example, storage capacity equals the MAR and evaporation loss is neglected.

Figure 7.10 indicates the frequency with which the annual flow of the river is less than specified values. The graph was prepared from the annual flow record of the Vaal river at Vaaldam. From the curve, it can be seen, for instance, that the probability that the annual flow will be between 50% and 100% MAR is 64-25 = 39%.

Divide the storage capacity into intervals 0–20%, 20–40%, 40–60%, 60–80%, 80–100% of the total capacity, and denote the probability of commencing a hydrographic year in the range 80–100% full by $P_{0.8-1.0}$ etc.

The following storage equation is used;

$$! = S_f - S_i + 0 \qquad\qquad (7.51)$$
where S_f = final storage
S_i = initial storage
0 = outflow

Several combinations of inflow and initial storage could result in the same range of storage at the end of a year. Thus storage state 80% – 100% could exist at the end of a year if any of the following occurred.

(i)   If storage state at the beginning of the year was 80% – 100% MAR and inflow equalled outflow: The mid-point of the range 80% – 100% is 90% which will be taken as the initial storage. The draft for a storage of 90% MAR is, according to Figure 7.9, 60% MAR. Inflow should exceed outflow by an amount equal to the draft less initial storage plus final minimum storage : 0.6–0.9+0.8 = 0.5 MAR. According to Figure 7.10 the probability of this flow being exceeded is 1.0– 0.24 = 0.76. The probability of two events occurring simultaneously, however, is equal to the product of the partial probabilities. It follows that the total probability of commencing a year with storage state 80 – 100% MAR and of inflow being such as to result in a final storage state 80 – 100% MAR is equal to $0.76 \times P_{0.8-1.0}$.

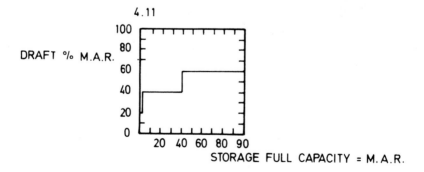

Fig. 7.9 Variable Draft Operating Rule

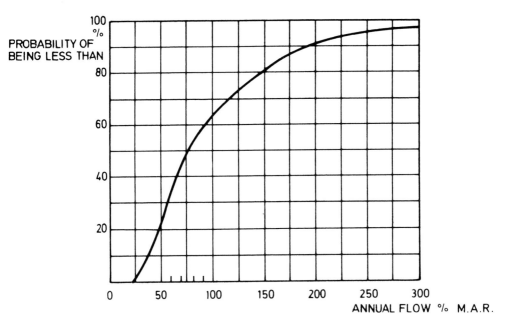

Fig. 7.10 Annual Flow-Frequency Curve

ii)   A storage state 80% – 100% MAR could exist if storage at the commencement of the previous year was 60 – 80% MAR, and inflow exceeded outflow by an amount equal to the draft less initial storage plus final minimum storage: 0.6-0.7+0.8=0.7 MAR. The probability that this flow will be exceeded is 1.0-0.46=0.54, so the probability of commencing a year with 60-80% MAR and ending with 80-100% MAR is 0.54 $P_{0.6-0.8}$.

The total chance of commencing a year with storage 80-100% MAR equals the sum of the probabilities of arriving at that state, from all possible storage states, at the beginning of the previous year. A calculation similar to those in steps (i) and (ii) is performed for each possible initial storage state. The coefficients thus calculated are summarized in Table 7.1 in the row coinciding with $P_{0.8-1.0}$. The numbers in that row are in fact the coefficients in the equation;

$$P_{0.8-1.0}=0.37P_0+0.33P_{0-0.2}+0.41P_{0.2-0.4}+0.41P_{0.4-0.6}$$

$$+0.54P_{0.6-0.8}+0.76P_{0.8-1.0} \qquad (7.52)$$

An equation may be derived for each final storage state. For example computation of the probability of starting with storage 0–20% MAR and ending with 60–80% MAR proceeds as follows:

According to Figure 7.9, draft for initial storage of 0–20% MAR is 40% MAR. The lower limit to the inflow is that which will raise the storage from 10% MAR (average initial value) to 60% MAR. Draft minus initial storage plus final storage equals 0.4–0.1+0.6=0.9 MAR. The upper limit to the inflow is 0.4–0.1+0.8=1.1 MAR. According to Figure 7.10 the probability that inflow will be in the range 0.9 to 1.1 MAR is 0.67–0.59=0.8. Similar coefficients are derived for different initial storages and assembled in the appropriate line of Table 7.1.

TABLE 7.1 Coefficients in the probability equations

Probability at the end of year	Probability at beginning of year					
	P_0	$P_{0-0.2}$	$P_{0.2-0.4}$	$P_{0.4-0.6}$	$P_{0.6-0.8}$	$P_{0.8-1}$
P_0	0	0.04	0	0	0	0
$P_{0-0.2}$	0.13	0.20	0.04	0.04	0	0
$P_{0.2-0.4}$	0.24	0.22	0.20	0.20	0.04	0
$P_{0.4-0.6}$	0.16	0.13	0.22	0.22	0.20	0.04
$P_{0.6-0.8}$	0.10	0.08	0.13	0.13	0.22	0.20
$P_{0.8-1.0}$	0.37	0.33	0.41	0.41	0.54	0.76
Answer	0.0003	0.006	0.036	0.096	0.194	0.668

The matrix in Table 7.1 represents a set of six equations in six unknowns. The simplest method of solving for the six probabilities appears to be by successive approximation. A first set of probabilities is arbitrarily selected such that their total equals unity. For example, the set of equations was solved by initially selecting the values 0.01, 0.03, 0.03, 0.08, 0.15, 0.7 for the columns in Table 7.1.

These probabilities were multiplied by the coefficients in Table 7.1 and each row summed to give a new value to the probability in the left-hand column. The new values were substituted into the equations to yield a revised set of numbers. After the third revision the values were found to be fairly stable. The final probabilities appear at the bottome of Table 7.1. An interpretation of the answer in the last column is that there is a

66.8% chance of commencing the year with storage in the range 80–100% full, if the reservoir is operated according to the rule depicted in Figure 7.9. Note that it is assumed a steady state has been reached, i.e. a number of years have passed since the reservoir was built.

The technique described could be used to derive an optimum variable-draft operating rule. If monetary values were assigned to the different levels of draft, the average annual economic loss contingent upon water rationing would be the sum of the probabilities of drawing water at various rates multiplied by the economic loss associated with the corresponding draft.

The optimum rule would be selected from a number of computations using alternative operating rules (Morrice and Allan, 1959).

A more direct method of deriving an operating rule that the foregoing would be to treat the draft at each storage state as a variable. The queuing equations would then include a term for each possible storage state and draft. The optimum draft at each storage level could then be selected by linear programming.

The method could be extended to permit a number of seasons to be examined instead of one whole year. The probability of starting any season with a certain storage level would equal the probability of ending the preceding season at that storage level. Recognition of a number of times of the year at which decisions might be made would enable the draft to be varied at more frequent intervals thereby rendering the system more flexible and possibly reducing the incidence of prolonged water rationings. In practice, however, it is unlikely that the effects of sequential correlation of monthly flows can be readily accounted for and the additional computations entailed in breaking the year into more than two seasons would probably not be justified.

Evaporation effects were omitted in the example, but could readily be taken into account in the release function in Figure 7.9.

REFERENCES

Alexander, G.N., 1962. The use of the Gamma distribution in estimating regulated output from storages. Civil Engineering Trans. Inst. Civil Engineers, Australia. CE4(1), 29–34.

Brittan, M.R., 1961. Probability analysis to the development of a synthetic hydrology for the Colorado River, in Past and Probably Future Variations in Streamflow in the Upper Colorado River, Part IV. University of Colorado.

Fiering, M.B., 1967. Streamflow Synthesis. Harvard Univ. Press. Cambridge, Mass., 139 pp.

Gringorten, I.I., 1963. A Plotting rule for extreme probability paper. J. Geoph. Res. 68(3) 813–4.

Gumbel, E.J., 1954. Statistical theory of droughts. Proc. ASCE 80. p. 439.

Gumbel, E.J., 1958. Statistics of extremes. Columbia Univ. Press, New York.

Haan, C.T., 1977. Statistical Methods in Hydrology. Iowa State Univ. Press. Ames. 378pp.

Hazen, A., 1914. Storage to be provided in impounding reservoirs for municipal water supply. Trans., ASCE, 77 No. 1539.

Hufschmidt, M.M. and Fiering, M.B., 1966. Simulation techniques for design of water resource systems. Harvard Univ. Press, Cambridge, Mass.

Klemês, V., 1979. Storage mass–curve analysis in a system analytic perspective. Water Resources Research, 15(2), p.359–370.

Langbein, W.B., 1958. Queuing theory and water storage. Proc. Amer. Soc. Civ. Engrs. J. Hydr. Div.

Maass, A. et al., 1962. Design of water resource systems Harvard Univ. Press, Cambridge, Mass., Ch. 14.

McMahon, T.A. and Mein, R., 1978. Reservoir Capacity and Yield. Elsevier, Amsterdam, 213pp.

Midgley, D.C., 1967. Operation of multi–unit–multi–purpose water development systems. International Conference of "Water for Peace", Washington, D.C.

Moran P.A.T., 1959. The Theory of Storage, Methuen, London.

Morrice, H.A. and Allan, W.N., 1959. Planning for the ultimate hydro development of the Nile Valley. Proc. Instn. Civ. Engrs. London. Vol. 14. p101.

Paling, W.A.J. and Stephenson, D., 1988. Prediction of cyclic rainfall and streamflow. Int. Wat. Ress. Assn. Conf. Rabat.

Rippl, W., 1882. Water supply of Vienna. Proc. Instn. Civ. Engrs. Vol. 71, 220.

Rippl, W., 1883. The capacity of storage–reservoirs for water supply. Min. Proc. ICE. LXXI; 270–278

Stall, J.B., 1962. Reservoir mass analysis by a low flow series. Proc. ASCE, 88(SA5), 3283, p21–40.

Stephenson, D., 1970. Optimum design of complex water resource projects. Proc. Amer. Soc. Civ. Engrs. 96 (HY6) p1229–1246.

Sudler, C.E., 1927. Storage required for the regulation of streamflow. Trans. ASCE, 91, No. 622.

Viessman, W., Knapp, J.W., Lewis, G.L. and Harbaugh, T.E., 1977. Introduction to Hydrology, 2nd ed. Harper & Row, N.Y. 704 pp.

CHAPTER 8

HYDROMETEOROLOGICAL NETWORK DESIGN AND DATA COLLECTION

INTRODUCTION

Developing countries frequently have poor or non-existent records. Hydrological data is a particular problem as fairly sophisticated technology is needed, not just instrumentation. That is, weirs need to be built, or river sections rated. Stage – discharge tables are needed, and these need periodic review as siltation or erosion can affect them. Channel roughness also changes with stage so that inaccurate flood estimates may result even though low flows are gauged accurately.

Often rainfall records are more comprehensive and reliable than stream flow records. Frequently, computer modelling may have to be used to obtain stream flow records. The accuracy of modelling is however limited so that a large inherent error in water resources planning may occur. The cost of over design or under design (Stephenson and Collins, 1988) can be severe, and can produce high project costs. Another problem is the impatience of designers. At least 10, preferably 20 years of reliable record is needed for major projects if risks are to be minimized, so planning of such projects should occur well ahead of construction. This chapter is intended to assist in establishing networks and collecting data in the most cost efficient manner.

NETWORK DESIGN (Clark, 1988)

Literature related to the design of hydrometerological networks is very extensive (Falkenmark, 1982). Several symposia have been convened concerning networks (e.g., International Association of Hydrological Science (IAHS), 1965 and 1986; American Geophysical Union, 1979). In 1972, the Casebook on Hydrological Network Design Practice (WMO, 1972) was published, and supplements added in 1978 and 1981. Rodda (1969) pointed out that the basic questions of station network-design are simple:

(a)   How many stations?
(b)   When?
(c)   For how long?

Those questions, are of course, an over-simplification since economic considerations and data quality are also of critical importance.

Data are the life blood of hydrological models and the heart of the systems which can simulate river flow conditions upon which good water management decisions can be made. Data may include rainfall, streamflow, temperature, humidity, snow water equivalent, snow depth, cloud cover, radiation, evaporation or evapotranspiration, groundwater levels and wind. Data requirements are dependent on the types of hydrological relationships utilized, area of concern and accuracy requirements. The time scale is also important since requirements may range from estimating flash flood peaks for forecasting or design to estimating annual runoff for water management or water supply. As pointed out by Langbein in the Introduction to the WMO Casebook (1972), even "the newer sophisticated planning and operational models have created a condition where the limiting factor continues to be data – data for evaluating the model, a different set for testing the model projections, and a subsequent set for auditing the performance of water projects."

Because of spatial variability, emphasis here in network design will be placed on precipitation and streamflow networks. Except in mountainous areas, temperature, evaporation, radiation and most other hydrometerological variables tend to vary slowly with distance, exhibiting primarily diurnal variation.

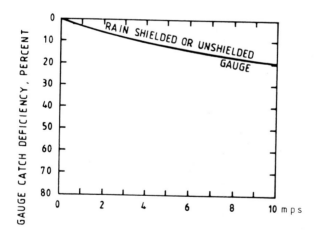

Fig. 8.1 Effect of wind speed on the catch of precipitation gauges. (Larson and Peck, 1974)

PRECIPITATION

Gauges – In general, the diameter of the rain gauge (i.e. from 12 mm to 200 mm) does not appear to be critical. However, gauge shape and exposure can produce different results because of varying wind and splash effects. The effect of wind speed on gauge catch is shown in Figure 8.1. Catch deficiences of 50 to 70 percent are possible during snow periods – even with shielded gauges.

The most common recording rain gauges are the weighing, tipping bucket and float gauges. The primary disadvantage of the tipping bucket and float type gauges is that any rain falling during movement of the control mechanism, i.e., during siphoning with the float type and tilting of the tipping bucket, is generally lost. These can produce errors of about 5 percent. However, the simplicity of the tipping bucket gauge makes it economical and easy to install and maintain.

The network density   is determined primarily by the uses for   which the data are intended. An early study of several storms in the Muskingum

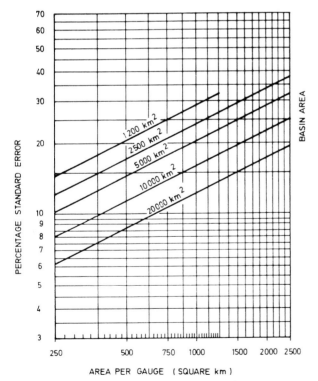

Fig. 8.2 Density-area-error graph (U.S. Weather Bureau, 1947).

Basin, Ohio, USA, shows that the standard error of rainfall averages for individual storm events varies with basin size and network density, (see Figure 8.2). It should be noted from Figure 8.2 that a basin area of 20 000 km² will produce an error of 10 percent with a gauge density of 700 km² per gauge (28 gauges) while a 5 000 km² basin must have a gauge density of 250 km² per gauge (20 gauges) to yield the same error. Thus it is apparent that the number of gauges required to estimate total or mean basin precipitation does not change significantly from small to large basins. These results are borne out by Johansen (1971). He explored the errors involved in simulating streamflow from a dense rain gauge network in central Illinois. Figures 8.3 and 8.4 indicate the calibration and dispersion errors resulting from a varying number of gauges over basins of various size. Figure 8.3 shows errors in estimating annual streamflow and Figure 8.4 the error in simulation of storm runoff.

The calibration error is the ratio (percent) of the standard error of estimate to the standard deviation of the observed flow. The dispersion error is expressed as a percentage of the coefficient of variation of the historic record. It should be noticed that neither error is closely related to network density. It is interesting to note that if an error of 10 percent can be tolerated, 4 to 6 gauges appears adequate regardless of basin size.

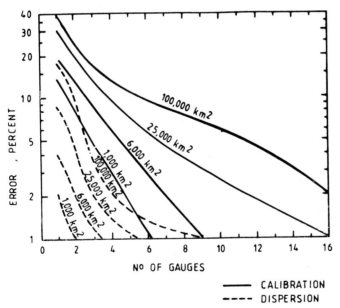

Fig. 8.3 Error of simulation of annual volume of streamflow for various number of gauges (Johansen, 1971).

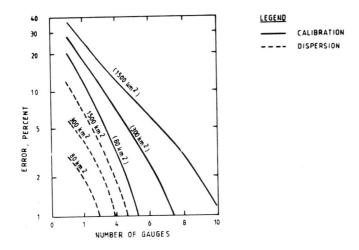

Fig. 8.4 Error of simulation of storm direct runoff for various number of gauges (Johansen, 1971)

Kohler (WMO, 1972) has developed a relationship between thunderstorm days and runoff per year, catchment area and the number of reporting precipitation stations necessary for flood forecasting, (Fig. 8.5). The standard error E (percent) from his study of average storm rainfall for a catchment area A in square kilometers and N stations is given by:

$$E = 7.7 \ (A)^{0.2}/(N)^{0.48}$$

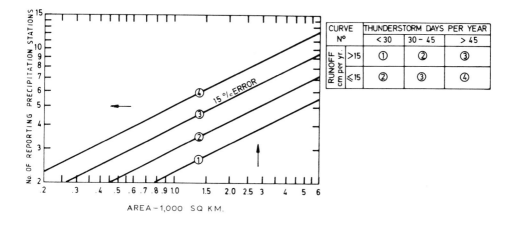

Fig 8.5 Minimum network requirement for continuous reporting stations (WMO, 1972).

Curve No. 3 in Figure 8.5 approximates a standard error of 15 percent in storm precipitation for regions experiencing 30 to 45 thunderstorm days per year. The other curves were based on judgment and experience. He also concluded that at least two stations for each area or sub-area forecast are a minimum requirement.

RAINGAUGES AND WEATHER STATIONS

The time resolution of data logging required to meet the objectives cited previously range from monthly averages on a basin scale to the order of minutes on catchment scale. Thus an integrated structure of both recording and non-recording instrumentation should be adopted.

Three levels of stations are proposed:

Level 1 – Weather Stations measuring the following parameters – autographic rain, totalizer rain, evaporation, wind direction and speed, humidity and where considered to be of use, sunshine recorder, atmospheric pressure.

Level 2 – Rainfall Recorder Stations – autographic raingauge, totaliser raingauge.

Level 3 – Raingauge Stations with totaliser gauge (manual collection).

Due to undular topography precipitation totals can vary considerably. Therefore the positioning of raingauges has to be undertaken not only with respect to climatic conditions and catchment usage but also considering topography. The guidelines of one autographic gauge per tertiary catchment and one non-recording raingauge per two quarternary catchments would also provide a baseline with which to work. A further aspect to be considered is the operation of these gauges, especially in remote areas.

A major constraint is the access to gauges. In addition, river basin and small catchment hydrology should be available. Hamlin (1983) states that for basin measurement of point rainfall the use of daily rainfall may be adequate, but for small catchments this will be insufficient, and the density of raingauges needs to be greater. For flood response and effect of urbanisation on small catchments, high resolution rainfall measurements are needed which require autographic gauges and accurate time correlation. The design of the network is seen as a first stage as it is important that the data from the network is assessed (Markham and Heyman, 1982) to improve the capabilities of the network. For specific areas where research needs are identified, further gauges will have to be placed. Clark (1988) suggests at least 5 gauges per catchment to be studied.

Bearing in mind that temperature changes mainly with elevation, and that evaporation does not vary as much across a country as does atmospheric pressure, wind and humidity, the number of such gauges may be minimised.

RECORDING APPARATUS

Present data capturing equipment often comprises clockwork driven paper charts with pens. These are subject to breakdowns and are tedious to process.

Electronic data loggers can capture and process data for new stations at a fraction of the cost. The use of data loggers offers a breakthrough in control of sampling and recording procedures as well as data processing. Such data loggers can capture a number of different items of information at specified time intervals or in analogue fashion continuously for a certain length of time. Built-in clocks ensure accurate synchronisation, and batteries can last a number of weeks or months. Batteries could be replaced when data is collected or re-charged or continuously charged using solar panels. At stream gauges additional channels could later be used for water quality monitoring. The data loggers can be designed to take samples at specific intervals or to take samples and readings when specified changes in values are exceeded. Loggers also provide checks to ensure that gauge observers actually attend the station regularly as they are not easy to update at a later time if an observation is missed.

The data in these loggers is captured by removing EPROM's (erasible programmable read only memories) or RAM's or by plugging in a portable computer or data capturer which is, after a data retrieval tour, returned to the central processor and plugged into any computer. The data is then stored on discs and can be processed easily. In fact this relieves the processor of the drugery of digitisation and enables a very broad perspective to be obtained. Summaries in various forms can be obtained at very short notice and patching or statistical analysis can be done with standard packages.

Micro-computers can be used for data processing, in which case hydrological staff should have access at all working times. IBM type PC compatibles can be used for most data loggers. At a later stage the meteorological and hydrological data can be synthesised and used in catchment models. These models can be used for predicting flows, statistical analysis to obtain probabilities of droughts or floods and for advance flood warning systems.

Many existing chart recording systems are currently operational, and the digitising of these charts can be done with computers and plotters in order to bring past records and continuing records up to date. Such programs are available, and graphic depiction of the digitised chart on the screen provides an easy check method. Rating curves can be programmed into the computer so that the digitised data, or in the case of logged data, the direct digital values will be translated to flow rates and separate tables of the flows will be maintained and updated using spread sheet methods.

JOB CREATION AND TRAINING

The expansion of hydrological networks provides a number of opportunities to advance employment. Not only will installation and weir construction provide employment, but in addition the data collection and processing will provide opportunity for development.

The following aspects can be incorporated:

i)   Labour intensive construction methods: Measuring weirs are generally stable and not hazardous structures so that quality of construction and timing are not critical. The weir thus provides ideal facilities for developing labour intensive methods and training schemes. Contracting type systems should be developed to encourage advancement of people with ability.

ii)   Field observers can be trained in hydrological techniques, for assisting in data collection and dissemination of data to farmers.

iii)   Technicians should be trained in data processing, digitising, computer usage and statistics to be able to interpret data..

STREAMFLOW

River Gauging

The objective in developing a stream gauge network is to improve the data available for water resources evaluation and flood estimation. A practical limit should be set on the number of proposed gauges, bearing in mind the difficulty of obtaining and processing many gauges and the cost of construction of weirs or even the equipment for measuring and recording water levels or flows. Gauge stations are spaced so as to obtain a reasonable indication of flows from various types of catchments, and some gauges are established specifically for selected rivers. Gauge stations must be sited bearing in mind access and the suitability of the river for either construction of a weir or for rating the cross section (which should be reasonably stable). A preference for weirs as opposed to rated natural sections is made where trained hydrologists for rating natural rivers are lacking and because weirs generally give more accurate data, especially at low flows. Weirs also provide employment during construction. Access is important for the purpose of retrieving data as well as construction. The

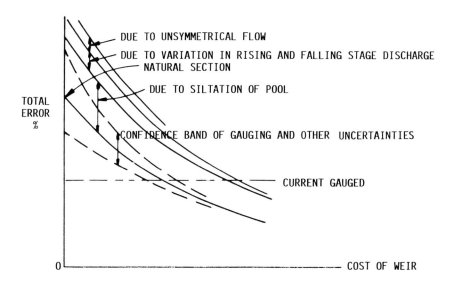

Fig. 8.6   Cost versus Weir Accuracy

170

decision as to whether or not to construct a weir should be based on the importance of the site, but also on whether a weir can be practically constructed, and on whether the section is stable and reasonably satisfactory for rating of the natural river cross section. Owing to the flat gradient of rivers near the coast, tidal effects and backwater rule out accurate gauging. The most important areas with respect to water utilisation are the interior region and the coastal escarpment region where there are possible hydroelectric sites. In general the sections should also be straight and uniform upstream, and a pool should be created by the weir to ensure low velocities over the weir. There should be no bends or obstacles downstream to cause backwater or non uniform flow profiles.

The retrieval and processing of the data are as important as the establishment of the gauging stations, and training programmes should be given for data collectors and processors.

Where data are to be digitized and fed into computers the software should be available and, in fact, the effort and cost can be significant. Data processing will make users aware of problems in data collection and make the data more meaningful to the processor when he sees summaries of flows.

Some gauging sites can be influenced by downstream conditions especially at high flows, and this and other effects can affect theoretical

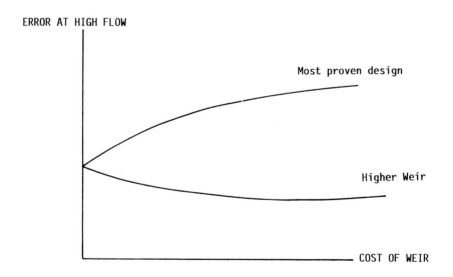

Fig. 8.7   Effect of Weir Construction on Gauging Accuracy

rating curves. For this reason it is advisable to calibrate some sections using both water profile computations and current meters. This can often be done from bridges, but in other cases a cable is required with bosun chair to traverse the section, especially during flood flows. Salt dilution methods are also possible but expensive as are radioactive tracer methods.

Weir Design (e.g. Ackers et al, 1978)

Older weirs are often sharp crested rectangular compound weirs. There is a problem of siltation upstream of weirs in some cases, however, due to high silt loads in many rivers this is difficult to avoid unless the river reaches are selected carefully or alternative designs are used. The Crump weir is now favoured for de-silting, but it requires a good foundation as it is very broad. Most river beds are alluvial and heavy grouting or excavation would be required; also the concrete volume required is considerably greater than for sharp crested weirs. There is also the problem of peak floods which will overtop any reasonable weir, and the actual design of the weir is not all that relevant during high flows because of inundation (e.g. Charlton, 1978), (see Figs 8.6-7).

In general, weirs are designed to monitor accurately flows up to approximately the 1-year flood without overtopping the flanks. Above this the cross section of the valley will also have to be considered, and rating curves may be required for estimating higher flows. Concrete weir sections are generally grouted or founded on bedrock as seepage under the structure and around the flanks in alluvial material can be a real problem. It is suggested that the bottom crest be a minimum height above bed level to avoid inundation or backwater effects during the 1-year flood flow through the section. Similar criteria should apply to each crest. The maximum rise of each step should be 400 mm except in a steep sided valley; then dividing walls should be used to ensure accuracy of the calibration curves. The crest and sides of each level should be mounted with 100 × 100 × 10 mm angle irons. Weirs should be located in reaches relatively straight upstream to ensure symmetrical flow. Relatively steep sections are preferred (BS, 1968) because:

a) They assist in suspending silt and scouring it out upstream of the weir.
b) They minimise the problems of backwatering and even tidal effects which inundate the weir.

Supercritical flow should, however, be avoided. An upstream pool for stilling flow is also required, although these frequently silt up and require recalibration of the weir.

The older type float operated water level gauges are reliable and robust but at the same time piezoelectric or similar electronic water level recorders are much cheaper. These are more economical than float systems also in data collection and processing, and more foolproof. Experience is, however, limited in electronic data collection, and it is suggested that one stage of such water level recorders be installed initially in order to prove their efficiency and discover the various problems. It is therefore suggested that provision be made for float chambers; i.e., 150 mm pipes should be constructed from just upstream of the lowest weir crest into a float chamber on the bank of the river. The chamber shaft should be constructed of masonry or other rigid material to avoid being washed away in floods and to a level which ensures that the housing of the recorder is not washed away with floods of less than, say, the 100 year flood (this will require a risk analysis).

The housings for electronic data collection systems can be simply enclosed boxes, but provision should be made for float and chart recorders (if required) together with their housing.

The rating of weirs can be carried out with the assistance of an available computer program, that will estimate backwater effect at each site for submergence of the weir. Each crest level can, thus, be selected to avoid submergence affecting flow significantly. Above the top of the weir the transition to full flood flow will be likewise estimated.

Gauges

The simplest form of a streamgauge is the staff gauge. Although they are easy to install, they must be read manually. Automated gauges are of several types – float type gauges requiring stilling wells and those utilizing some other type of device which measures water pressure (e.g. nitrogen bubbler gauges and other pressure measuring devices). In recent years many of the automated gauges have been equipped with various types of communication equipment (viz., telephones and radio utilizing both line of sight and satellite relay). All of the above only measure water stage and rely on a stage-discharge relationship to convert recorded readings to streamflow.

Networks

Most streamflow networks tend to be developed for specific purposes and, frequently, appear to have little relevance to national concerns with water. For example, the networks for flood forecasting normally include gauges primarily at forecasting points (usually locations with high potential flood losses) and may not be appropriate for irrigation, municipal water supply, industrial uses or navigation. Kovacs (1986) has demonstrated that runoff variability tends to increase with decreasing basin size. Also, the sampling interval is critical in determining data accuracy. Figure 8.8 relates drainage basin area to sampling interval. Nevertheless, for areas larger than 2 000 km² sampling should probably be at least twice each day since the error can become fairly large in estimating daily runoff.

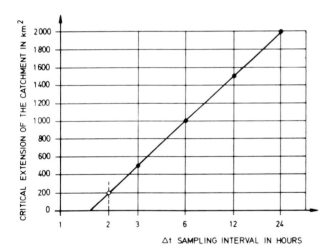

Fig. 8.8 Relation between a reasonable sampling interval and drainage basin area (Kovacs, 1986).

Moss (1979) has discussed a useful categorization in network design which employs a use-oriented taxonomy as follows:

1. project operations,
2. project design,
3. resource planning,
4. water policy development and evaluation, and
5. research.

Except for research, the above list is ordered by the general intensity of data demands; project operation generally requires more information than project design or resource planning. One of the primary purposes at many stream gauging stations today is related to water quality monitoring, e.g. in Kansas, USA, 79 of 119 stations (of course, other related uses are included at the same gauge – water availability, flood hazards, etc.)

The WMO has recommended that minimum gauging densities include:

1. For temperate and tropical climates, plains regions. 1 gauge per 3 000 to 5 000 km^2

2. For mountainous basins in temperate and tropical zones. 1 gauge per 1 000 km^2

3. For desert regions and polar zones. 1 gauge per 1 000 km^2

These densities are, of course, very tentative in that all networks are also subject to economic considerations and specific data requirements.

OTHER NETWORKS

Types

1. water quality
2. sediment measurements
3. groundwater monitoring
4. atmospheric wind and temperature
5. soil moisture
6. evaporation and evapotranspiration
7. water temperature and
8. ice on rivers and lakes

Networks for a few gauge types will be discussed briefly below. Most of these networks are of a special type and network densities vary considerably – depending on data requirements, e.g. groundwater networks are frequently very dense in areas of extensive pumping.

Groundwater

Because of a slow change in both quality and quantity, groundwater network problems are somewhat different than those in surface water hydrology. A major problem is that direct measurments are primarily limited to wells – which are expensive to construct. Time series data include:

1. water levels
2. volume of water withdrawn
3. volume of water coming from springs
4. volume of water discharged into the groundwater through wells or infiltration ponds and
5. quality of groundwater

The spatial variability is strongly dependent on the physical properties of the aquifer and on the hydrological regime. In some areas the density of well observations is very high, e.g. 228 wells in 100 km^2 in the coastal plain of Israel. In the United States the Geological Survey monitors more than 30,000 wells over 7,500,000 km^2.

Sediment

Sedimentation embodies erosion, entrainment, transportation, deposition and the compaction of sediment. All of these are natural processes caused by rainfall, runoff, streamflow and wind forces. In order to do a good job of measuring sediment, a thorough understanding of the above processes is necessary. Sediment transport is mainly made up of suspended solids and bed load, (Table 8.1) controlled primarily by velocity and particle size.

Obviously, in regions of large sediment movement, sediment measurements need to be made at almost all streamflow stations.

Discussion of sediment measurement samplers and sampling is given in a report edited by Vanoni (1975). Since the concentration of sediment is highly correlated with discharge, during periods of high flow samples need to be taken frequently. For some flood events the peak sediment concentrations sometimes lag the peak flow which may alter a normal "load-discharge" relationship. Thus, measurements need to be made both before and after the flood peak. In areas where the sediment load is made up primarily of fine materials such as silt or clay (wash load), the transport curves may vary somewhat from those for sand.

TABLE 8.1   Kinds of Sediment

Sediment	Size class	Mode of transport
Boulders	Larger than 256mm	Bedload
Cobbles	64–256mm	Bedload
Gravel	2–64mm	Bedload
Sand	0.062–2mm	Bedload or suspended
Silt	0.004–0.062mm	Suspended
Clay	0.0002–0.004mm	Suspended
Organic detritus Includes leaves, trees, biologic remains etc		Bedload or suspended
Biota Includes bottom dwelling organisms		Bedload or suspended

Note: The size classes of mineral sediments shown are based on the Wentsworth scale (Lane and others, 1947).

Bed load estimates are somewhat more difficult to make than suspended load. The efficiency of samplers tends to vary considerably with hydraulic conditions, particle size, bed stability and bed configuration.

In addition to sampling sediment transport in streams, surveys of reservoir sediment deposits are also very important. The frequency of surveys depends upon the rate of accumulation in the reservoir. Surveys at intervals of 5 to 10 years are generally recommended. Following major floods, it is frequently important to make a survey.

It is difficult to define a sediment network. Normally, sediment and other parameters are measured at stations operated for streamflow since it is the primary causative factor to which all data can be related. Estimates of sediment transport cannot be made accurately with only one or two years of data; most hydrologists estimate a minimum of 10 to 20 years is required (10 years in wet climates to 20 years in arid regions). Data should be collected at sites where no major engineering works will be started during the period of study.

Water Quality

Networks for the measurement of water quality are similar to sediment networks – they are required for a special purpose. If the primary objective is loosely defined "to obtain information on the water quality of rivers," the information may be almost useless because:

1. The quality parameters must be specified,
2. The required sensitivity and accuracy needs to be stated,
3. A time scale and sampling frequency are needed,
4. The quality measurements must be expressed in selected terms (for example, averages or median values) and the tolerable uncertainty stated and
5. The use must be specified.

Technology available today permits very accurate measurement of water quality parameters – to parts per billion or trillion in many cases. Most water quality standards today are technology based, i.e., they are established by techniques available for measurement. In many instances they are based on known quality levels that are harmful to human consumption.

It is obvious that the network for sampling will normally be established as a result of program objectives. The points within the rivers at which samples are collected are also important. Uniformity with which the impurities are distributed throughout the river cross section must be checked.

The time of sampling also varies considerably. Large variations in water quality loading occur throughout the day, month and year. It is wise under such circumstances for the samples to be taken using known statistical sampling techniques since the results can be easily biased.

Evaporation

There are a number of ways of measuring evaporation – the simplest form of data is obtained from evaporation pans or tanks. Unfortunately, most pans must be multiplied by a factor (0.7 for the U.S. Class A pan) to convert the data to lake or reservoir evaporation. Other more refined techniques include the lysimeter, heat budget, mass transfer and water balance methods. All of these require extensive and expensive instrumentation.

Estimates of evaporation can also be made using a number of empirical formulae. These normally require extensive meteorological instrumentation.

Fortunately, the areal variability of evaporation is much less than that for rainfall. Also, temporally, evaporation tends to follow a daily and seasonal variation. It is, thus, not necessary to employ the network densities and data collection frequency required for either rainfall or streamflow.

Studies made in the United States have indicated a network density of about one evaporation pan for each 15,000 km^2. Pans are frequently installed at most major reservoirs. Normally at such sites, rainfall temperature, wind and humidity data are collected also.

REMOTE SENSING

In many remote areas, installation of gauges and data processing may not be justified or even possible. Then more general methods of obtaining data may be necessary.

Remote sensing is defined as the observation of a target by a device at some distance and may be either active or passive sensing. Active sensing includes radar which measures back-scattered microwave energy – primarily valuable for estimating rainfall and tracking severe weather systems. Passive sensing includes electromagnetic radiation reflected or emitted by the earth or atmosphere. From the measurement of intensity and spectral distribution, several hydrometeorological parameters can be estimated – e.g. soil moisture, snow water equivalent, snow cover, precipitation distribution, area extent of floods, land use, surface temperature, evapo-transpiration/evaporation and geographic information. The electromagnetic spectrum is presented in Figure 8.9. Most passive sensing involves the infrared portion of the spectrum. Recently, considerable interest has been shown in the microwave portion of the spectrum from about 0.1 to 50 cm and in the wavelengths below the visible region – less than 0.4 microns.

Passive sensing can be accomplished using satellites or from aircraft. In recent years low-flying (100m) aircraft have been used to sense radiation in the Gamma Ray portion of the spectrum to measure soil moisture and snow water equivalent. Also shown in Figure 8.7 are portions of the electromagnetic spectrum experiencing strong atmospheric attenuation. It is apparent that microwave wavelengths are advantageous for remote sensing (both active and passive) because of minimal attenuation.

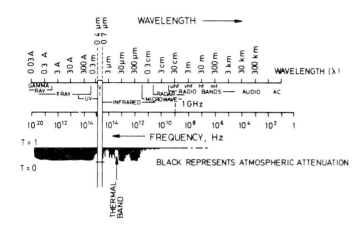

Fig. 8.9 Schematic representation of the electromagnetic spectrum. The
bottom figure is atmospheric transmissivity as a function of
frequency (Schmugge, 1985).

Radar

Radio detection and ranging (RADAR) was first used 50 years ago to
detect aircraft. It soon became apparent that clouds with precipitation
sized droplets (greater than 0.1mm) were also detected easily by radar
operating in commonly used wavelengths. Radar is an active sensor in that
energy is transmitted in short bursts and the back-scattered energy from
particles of ice, snow or water droplets is detected. The strength of the
received power varies with the electromagnetic properties of the particles,
the number and size of the particles, refraction in the intervening medium
and atmospheric absorption (mainly by oxygen and water vapour). The
basic radar equation for computing the back-scattered power, Pr, is:

$$Pr = \frac{C \, |K|^2 \, Z}{r^2}$$

where C is a constant for any given radar and a function of antenna size
and shape, power transmitted, wavelength, pulse length and other
hardware characteristics, K is a dielectric factor related to whether the
scatterer is ice or water ($|K|^2$ is approximately 0.93 for water and 0.197
for ice), Z is a reflectivity factor and equal to the sum of the drop
diameters (D) to the sixth power ($Z = \Sigma D^6$), and r is the distance to the
target.

Two important factors enter into determination of the magnitude of the received power – $|K|^2$ for water is about 5 times larger than for ice, and, since Z is related to the sixth power of drop diameter, a 1 mm drop (light rain) will return 10^6 more power than a 0.1 mm drop (drizzle). Another factor not indicated above is that C (radar constant) is inversely related to the wave-length of the radar to the second power. Thus a 3 cm radar can detect signals 10 times weaker than a 10 cm radar. Unfortunately, 3 cm radiation is also severely attenuated by moderate to heavy rainfall – making estimates of rainfall unreliable. Figure 8.10 shows estimates of rainfall rates using 3 different wavelengths. The 3.2 cm estimates are very low while those based on 10 cm radar are fairly reliable. This large variation is caused primarily by severe attenuation of the 3.2 cm and 5.5 cm wavelengths.

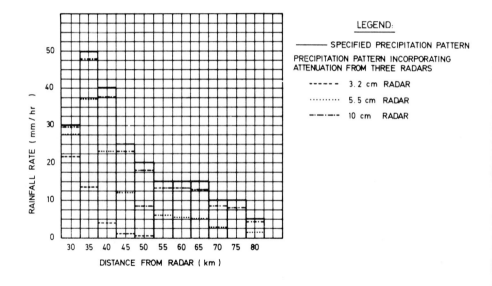

Fig. 8.10 Precipitation estimates based on attenuation and back scattering of three wavelengths (Huebner and Leary, 1982).

Rainfall intensity is, of course, related to the drop size. If one assumes no vertical air motion (obviously an incorrect assumption in a severe storm system), the drops can be assumed to be falling at terminal velocity. A relationship can be developed then between drop diameters (third power) and rainfall intensity. Thus, the following equation relating Z and R can be approximated:

$$Z = AR^b$$

where b under ideal conditions would be about 2, but normally is around 1.6. The constant A is frequently assumed to be about 200 – actually varying widely from 55 to 1 000.

It should be apparent from the above discussion that rainfall intensity estimates based on radar detection alone can produce widely varying results (easily a factor of 2 to 4 between radar estimated rainfall and observed rainfall). The capability of radar to detect a large area (200 km radius) instantaneously and frequently (at least 3 full sweep scans per minute) provides a powerful tool to detect not only where it is raining but whether the rainfall intensities are varying in both a spatial and temporal sense. Each digit is related to power received or rainfall intensity. Recent studies have indicated that reasonably accurate calibration of radar estimated rainfall can be achieved using a few automated rain gauges located under the radar umbrella.

Satellites

Both polar orbiting (NOAA, TIROS) and geostationary (GOES, METEOSAT, GMS) satellites are being used extensively today in hydrology and meteorology. The polar orbiting satellites operate at much lower elevations, 1 000 km versus 36 000 km for geostationary satellites, and are capable of measuring much smaller surface elements, 30 m versus about 1 100 m. The polar orbiting data are only available over a specific point about once each 18 days while geostationary data can be obtained each 30 minutes. Data are normallly collected in the visible and thermal infrared bands.

Satellite information has proved valuable in estimating rainfall intensities and areal extent. Figure 8.11 illustrates procedures used in satellite rainfall estimation. Basically, experience has shown that rainfall intensities are related to the height of cloud tops. The top elevation can be estimated by estimating top temperatures using infrared imagery. The intensity of radiation in the infrared region is related to the temperature of the radiating body to the fourth power (Stefan–Boltzmann equation).

A comparison of isohyetal patterns from satellites and ground truth is presented in Figure 8.12.

Extensive use has been made of satellite estimation of snow cover extent. In many areas of the world, e.g., western United States, the Alps in Europe and the Himalayan Mountains of Asia, 50 to 90 percent of the annual total runoff is produced from snow melt. The primary problem is that cloudy conditions frequently preclude observing snow cover in both the visual and infrared wavelengths.

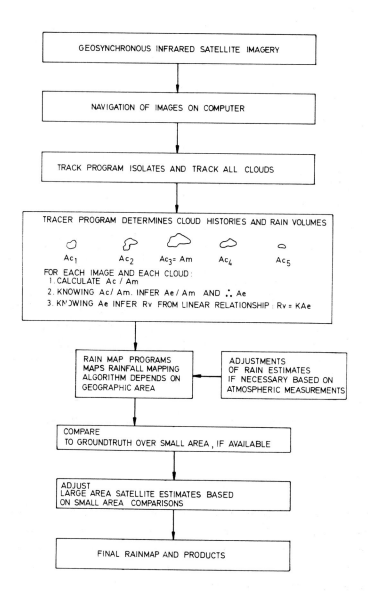

Fig. 8.11 Schematic of the steps involved in rainfall estimation
by satellites (Woodley, et al., 1981).

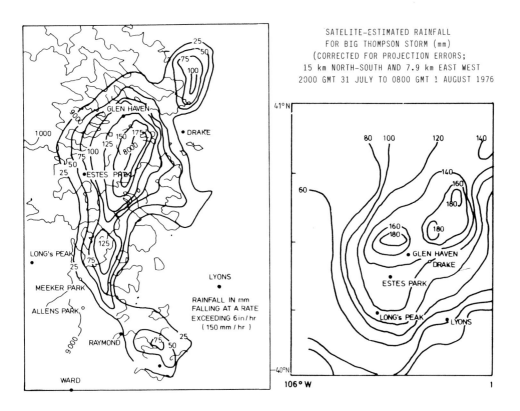

Fig. 8.12 A comparison of isohyetal patterns from satellites (right) and ground truth (left) (Caracena, et al., 1979).

Microwave Radiometry

Reasonably accurate measurements of soil moisture can be obtained by measuring the intensity of thermal emmission from the surface at wavelengths in the microwave region. Fig. 8.13 is a schematic diagram of this type of measurement. Unfortunately, at elevations in which most polar orbiting satellites operate (greater than 500 km) fairly large antennas (10 m) are required in order to obtain a spatial resolution of 10 m at a wavelength of 21 cm – an accuracy necessary to resolve soil moisture variations.

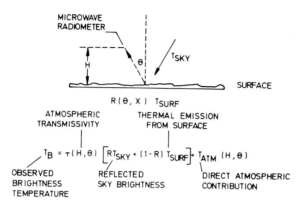

Fig. 8.13 Schematic diagram of the source of microwave radiation measured by a radiometer (Schmugge, 1985).

In Figure 8.14 the equation in Figure 8.13 is used to compute the brightness temperature T_B based on the surface albedo r, sky temperature T_{SKY}, atmospheric transmissivity and a direct atmospheric contribution T_{ATM}. Depending upon soil type and soil moisture content, the normalized value of T_B decreases with increasing moisture in almost a linear manner.

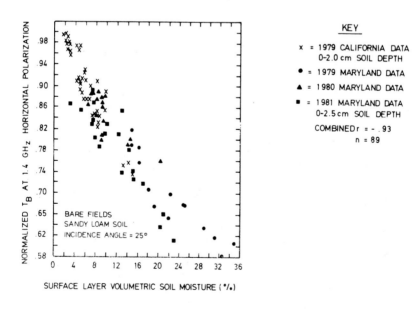

Fig. 8.14 Measurement of normalized T_B made from a tower, 21 cm wavelength (Schmugge, 1985).

Other Applications

In recent years the National Weather Service in the United States and the Hydrometeorological Service in the Soviet Union have employed low-flying aircraft (100 m) to measure natural gamma radiation from the earth's surface to estimate snow water equivalent and soil moisture.

Gamma radiation (well below the visible region, see Figure 8.9) is severely attenuated by liquid or frozen water. Measurements are usually made along a 15 km flight path, which is easily identified and can be repeated under various seasonal conditions. Using occasional ground truth observations to check on the calibration, it is possible to determine variations in the measured radiation intensities under moist conditions and when attenuation is caused by snow water. Accuracy of measurement is generally within 0.5 cm.

REFERENCES

Ackers, P., White, W.R., Perkins, S.A. and Harrison, A.J.M., 1978. Weirs and Flumes for Flow Measurement. John Wiley.

American Geophysical Union, 1979. Chapman conference on hydrologic data networks, Water Resources Research, 15 (6), 1673-1871.

British Standard 3680, 1986. Measurement of liquid flow in deep channels.

Caracena, F., Maddox, R., Hoxit, L.R., and Chappell, C.F., 1979. Mesoanalysis of the Big Thompson Storm. Monthly Weather Review. Vol. 107, No. 1, pp 1-17.

Charlton, F.G., 1978. Measuring flows in open channels, Report 75, CIRIA, London.

Clark, R.A., 1988. Design of hydrological networks. Course on Water Resources in Developing areas. University of the Witwatersrand, Johannebsurg.

Falkenmark, Malin, 1982. Water data strategy - A general approach, in Water for human consumption, man and his environment, IVth World Congress of the International Water Resources Association, Bueno Aires, Argentina.

Hamlin, M.J., 1983. The significance of rainfall in the study of hydrological processes at basin scale. J. Hydrol. 65. p 73-94.

Huebner, G.L. and Leary, C.A., 1982. Radar meteorology class notes. Texas A & M University, College Station, and Texas Tech University, Lubbock.

International Association of Hydrological Sciences (IAHS), 1965. Design of Hydrological Networks, Two Volume proceedings, IAHS Publication Numbers 67 and 68.

International Association of Hydrological Sciences, 1986. Integrated design of hydrological networks. Proceedings of the Budapest Symposium, IAHS Publication Number 158.

Johanson, R.C., 1971. Precipitation network requirements for streamflow estimation. Stanford Univ., Dept. Civ. Eng. Tech. Rep. 147.

Kovacs, G., 1986. Time and space scales in the design of hydrological networks, Integrated design of hydrological networks. Proceedings of the Budapest Symposium, IAHS Publ. No. 158. 283-294.

Lane, E.W., 1947. Stable channels in erodible material. Trans. Am. Soc. Civ. Engs. Vol. 102.

Larson, L.W. and Peck, E.L. 1974. Accuracy of precipitation measurement for hydrologic modeling. Water Resources Research, Vol. 10, No.4, pp.857–863, August.

Markham, R. and Heyman, C.A., 1982. A simulation study of techniques for annual rainfall estimation and the design for raingauge network, Natl. Inst. for Math. Sciences, CSIR, Pretoria.

Moss, Marshall, E., 1979. Some basic considerations in the design of hydrology data networks, Water Resources Research, Vol. 15, No. 6.

Rodda, J.D., 1969. Hydrological network design – Needs, problems and approaches. WMO/IHD Report No. 12, WMO, Geneva, Switzerland.

Schmugge, T., 1985. Remote sensing of soil moisture, Chapter 5 in Hydrological Forecasting. M.G. Anderson and T.P. Burt, eds., John Wiley & Sons.

Stephenson, D. and Collins, S., 1988. Problems due to innacurate flood estimates at Collywobbles. Proc. Intl. Congress on Large Dams, San Francisco, Q63V4, p693–700.

U.S. Weather Bureau, 1947. Thunderstorm Rainfall. Hydrometeorological Report No. 5. With the Corps of Engineers, Vicksberg, Mississippi.

Vanoni, Vito A. (editor), 745 pp. 1975. Sedimentation engineering, ASCE–Manuals and reports on engineering practice – No. 54, American Society of Civil Engineers, New York.

Woodley, W.L., Griffith, C.G., and Augustine, J.A., 1981. Rain estimation over several areas of the Globe using satellite imagery. Satellite Hydrology, American Water Resources Association, pp 84–91.

World Meteorological Organization (WMO), 1972. Casebook on hydrological network – design practice, WMO Publication number 324, Geneva, Switzerland.

CHAPTER 9

SOIL EROSION AND SEDIMENTATION

INTRODUCTION

A serious problem exists in many developing parts of the world particularly where agriculture is expanding. The problem under reference is rural resource degradation and more particularly soil erosion.

The problem of soil erosion in developing countries is ubiquitous and very serious and differs only in degree from place to place. The problem is far worse than is generally realised and the situation appears to be deteriorating. It has enormous social, economic and political implications and its impact on the more developed sectors can only be substantially adverse. The problem is inextricably linked with development and education. Orthodox engineering aspects are but a minor facet.

The Main Causes (Venn, 1988).

The dominant basic cause, and the ubiquitous cause, of soil erosion is excessive pressure on the resource base in a milieu of ignorance, poverty and land use malpractices.

The dominant specific cause of the worst soil erosion is generally vegetation denudation resulting from over-grazing as aggravated by the communal grazing system. (Every stockowner for himself and devil take the hindmost: the eventual result is BURBAR terrain). Note that technically, the matter, being essentially related to plant cover, is a biological as opposed to an engineering problem.

Attitudes to natural resources and agriculture aggravate the problem. Erosion is not regarded by most rural people as an important issue. National or tribal will to conserve natural resources is generally minimal. Generally the leadership, so influential in other fields, pays lip service to the conseration ethos. Complicating factors are the important social role of iivestock coupled with the economic sense of investment in livestock. There is also a general lack of understanding of land management and the reasons for degradation.

Facts

The problem is worst in the medium rainfall areas (400-600 mm p.a.)

that are marginal for crop production because (a) reduced carrying capacity is not matched by reduced livestock numbers and (b) the instability of many arid zone soils. Available land is finite in extent. The number of landless is growing. Populations will approximately double in 20 years. This is a useful time frame for planning.

The dominant need felt of most rural communities is improved water supplies. Fuel is a matter of growing importance as trees are cut down and dung is used instead, thereby further reducing fertility and vegetation.

Rates of erosion from farmed lands are summarized by Zachar (1982). He indicates rates as high as 500t/ha/a which corresponds to a depth of 50mm/a from poor grassed arid areas. The rate drops to less than 100t/ha/a for cropped land with slope less than 10% and even less than 1t/ha/a for well tended fields.

If these rates are compared with the regenerative capacity of the soil, they are indicative of large scale dissappearance of workable land in less than a century. Rates of soil formation can be 0.1 to 10t/ha/a or 0.01 mm to 1 mm depth/a. Generally soil reformation is more rapid in shallow soils, but these tend to be more erosive because they are probably on steeper slopes or more poorly managed. Regeneration may be due to weathering, or deposits by wind or water on flatter land.

TRAINING ASPECTS

There are few colleges or schools equipped or even a syllabus designed to teach conservation. For poor, ill educated people conservation is virtually meaningless except in the context of personal direct benefit. Therefore conservation programmes must be integrated with suitable development programmes. Rural people must be substantially involved in rural conservation and development planning and implementation. Conservation and development progammes must as far as possible (a) counter the fundamental causes of problems and (b) manifest attention to community needs.

The following methods do not appear to have worked:

Externally imposed discipline is required (e.g. paddocking, rotational grazing, stock reduction).

Appeal to the emotions ("conserve for future generations") or appeal to logic ("if you do not conserve your grass and soil your livestock will die in times of drought").

Widespread reform in land tenure. This indeed is part of an ultimate solution but under present circumstances is a pipe dream.

Any substantial conservation proposal that does not directly benefit the local people.

Old style "development work" based on engineering works planned and constructed by outsiders that today litter the sub-continent as monuments of failure.

There are no simple solutions to the problems of resource degradation in developing areas. Solutions to these problems will always be complex, and must be multi-faceted and holistic and deal with fundamentals.

The dimensions of these problems are such that there can be no total solution in our time. Constraints of time, money, manpower and others preclude this. There is not yet a shortage of land or food on an international scale, and world attention has therefore not focussed on the problem. The fact is that the problem will surface with a more abrupt jar than many environmental problems receiving the attention of 'Greens'. It is necessary to initiate integrated conservation and development programmes in selected small areas in the first instance and expand outwards thereafter. Education and training are key factors in the solution.

Important Facets in the Solution

By promoting urbanisation, the pressures on the land will reduce. By embarking on multi-facet programmes of land use planning for conservation and development in selected areas, preferably at the head of catchments, in the first instance, the erosion will reduce.

Having regard to the facts that : (a) water is a major cause of erosion, (b) typically, improved water supplies are the dominant felt need of rural communities in developing areas and (c) the favourable cost benefit ratio of innovative water based appropriate technologies, one could integrate soil and water conservation in research, planning and implementation and implement on the basis of labour intensive systems.

TABLE 9.4 Rivers of the World Ranked by Sediment Yield

(Source: Holeman, Water Resources Research, 1968. Copyright by Am. Geophysical Union)

River	Drainage basin. $10^3 km^2$	Average annual suspended load		Average discharge at mouth, $10^3 cfs$
		Metric tons $\times 10^6$	Metric tons/km^2	
Yellow	673	1 887	2 804	53
Ganges	956	1 451	1 518	415
Brahmaputra	666	726	1 090	430
Yangtze	1 942	499	257	770
Indus	969	436	449	196
Ching	57	408	7 158	2
Amazon	5 776	363	63	6 400
Mississippi	3 222	312	97	630
Irrawaddy	430	299	695	479
Missouri	1 370	218	159	69
Lo	26	190	7 308	–
Kosi	62	172	2 774	64
Mekong	795	170	214	390
Colorado	637	135	212	5.5
Red	119	130	1 092	138
Nile	2 978	111	37	100

Few universities produce the product we need. The developing regions need a large number of Conservation and Development graduates and technicians whose training should stand on four legs: e.g. elements of agriculture, engineering, sociology and economics.

One should enable local people to make beneficial use of the products of erosion e.g. "farm a donga" and/or store water in it; produce construction materials, fuel and fodder in silt deposits, and so on. It is necessary to undertake more research into social and economic costs and benefits of soil erosion and conservation, a field to which orthodox evaluations are difficult to apply.

RESERVOIR SEDIMENTATION

Construction of a dam and reservoir on a river interferes with stream equilibrium by modifying streamflow and, consequently, the sediment transport capacity of the stream. All reservoirs are subject to some degree of sedimentation (sediment deposition), and eventually all reservoirs will fill with sediment. Some will fill faster than others, depending primarily on sediment yield of the tributary drainage basin, transport capability of the stream, and size of the reservoir.

In water resource planning the problem is to estimate the sediment yield of the drainage basin, the rate and amount of sediment deposition in the reservoir, and the length of the time before deposition enroaches on the useful storage space in the reservoir to the point where it interferes with the system operating as it was designed to operate. In planning and design of reservoirs, it is essential that potential problems associated with sediment be considered. Sufficient storage must be provided so that sediment deposition will not impair reservoir operation during the useful life of the project or the period of time used in economic analysis.

Basin Sediment Yield

Sediment is derived from erosion (wearing away) of the land surface by natural forces – water, wind, ice, and gravity. Not all eroded material enters the drainage system, but what does is termed the sediment yield of the basin. Sediment transported by streams is derived from scour and erosion of the streambed and banks as well as from erosion of the land surface and rills of the drainage basin. Sediment yield varies with land slope, land use, vegetative cover, soil type, amount and type of precipitation, climatic factors, and nature of the drainage system. Natural erosion rates are accelerated by human activities, including deforestation, urbanization, farming, grazing and channelization of streams.

Worldwide, the annual sediment yield from drainage basins is highest in southeast Asia, the southeastern United States, and in the tropics, as shown in Fig. 9.1. Loess deposits, as in central China, also have a very high sediment yield. High latitude areas, with low precipitation and low runoff, have low sediment yield. In the mid-latitudes (55° - 30° N and S) and in the tropics, vegetation reduces surface erosion (Fig. 9.2). Areas with a marked dry season have high sediment yields because desiccation of grassland produces much erosion in the early part of the wet season (Petts, 1984).

CLASS	SEDIMENT YIELD $(t.km^{-2}yr^{-1})$	RUNOFF (mm)
1	ARID	< 50
2	0 - 50	50 - 500
3	0 - 50	500 +
4	50 - 100	50 - 500
5	50 - 100	500 +
6	100 +	50 - 500
7	100 +	500

Fig. 9.1 A general classification of world rivers. Sediment yield is indicated in tonnes per square kilometre per year (Petts, 1984).

In tropical and semi-tropical areas there is usually a distinct rainy season, lasting several months, and lesser rainfall throughout the remainder of the year; in such areas sediment yield is moderate to high, as in central Africa. Walling's (1984) estimate of suspended sediment yield on the African continent is shown on Figure 9.2. Runoff from thunderstorms carries larger concentrations of sediment than runoff from general rains.

Two processes are involved in soil erosion from land surface: sheet erosion and rill erosion. The intensity of sheet erosion is a function of: (1) Geomorphological characteristics of the basin (soil erodibility, steepness of terrain, and length of slopes); (2) Soil types; (3) Land use and agricultural practices; and (4) Climatic and precipitation factors. Erosion from rills is a function of: (1) Rill characteristics; (2) Seepage forces that may cause sloughing of rill borders; (3) Amount and type of clays in the soil; (4) Amount of organic material in the soil; (5) Size of soil particles; and (6) Climatic and precipitation factors.

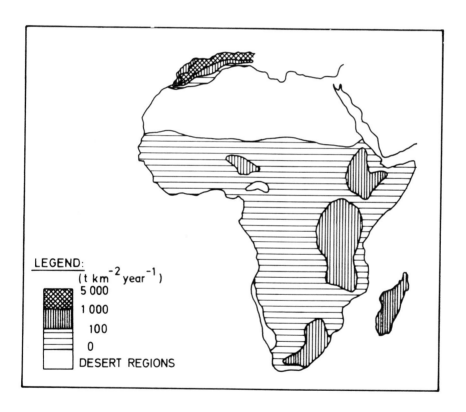

Fig. 9.2 A tentative and generalized map of the pattern of suspended sediment yields within the African continent. (Walling, 1984).

Schumm (1977) divides the fluvial system into three zones that serve as primary areas of sediment production, transfer, and deposition, respectively, as shown in Figure 9.3. Zone 1 is the drainage basin that is the predominate source of sediment and water. Zone 2 is a transfer zone for sediment from source area to point of deposition. If there is no significant trubutary inflow downstream of Zone 1 and if the channel is stable, sediment inflow at point A will equal outflow at point B in the figure. Zone 3 is an area of sediment deposition, such as an alluvial plain, alluvial fan, inland delta, or estuarine delta.

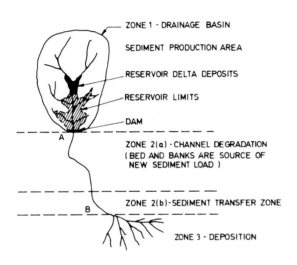

Fig. 9.3 Fluvial system with dam and reservoir

Effects of Impoundments on Sediment Transport

Streams transport sediments as both suspended and bed load. Where a river flows into a large body of water, such as a reservoir, the water depth and cross-sectional area increase and stream velocities decrease rather rapidly, thus reducing the sediment transport capacity of the stream and resulting in deposition of sediment in the headwaters of the reservoir. With time, some of the fine deposits move down through the reservoir and deposit against the dam, and some are flushed through the system. Typical deposition patterns are shown in Figure 9.4.

Most fine suspended load will pass through a small reservoir with a short detention time or a run-of-river low-head reservoir. However, larger reservoirs that impound water for long time periods can be expected to trap much of the suspended load as well as the bed load. The percent of inflowing sediment deposited in a reservoir (the trap efficiency of the reservoir) is a function of the ratio of reservoir capacity to total inflow.

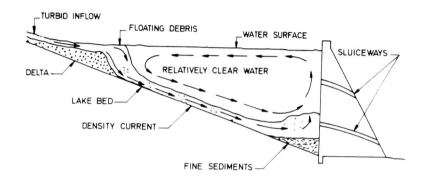

Fig. 9.4 Sediment Accumulation in a Typical Reservoir

Releases from a dam usually carry relatively little sediment as a result of deposition in the reservoir. Downstream from the dam the stream will pick up sediment from the bed and banks until it regains a normal sediment load; degradation occurs and river slopes decreases progressively in a downstream direction until the limiting conditions for sediment tranport are reached. If material comprising the channel boundary has a wide size gradation, natural sorting may result in the formation of an armor layer of coarse material that limits the extent of scour. The degradation process begins immediately below the dam and proceeds in a downstream direction until the equilibrium sediment load for local slopes and velocities is reached (Vanoni, 1946).

For the case of a dam and reservoir, the zones identified by Schumm for a fluvial system can be modified and defined as shown in Figure 9.3. Zone 1 is the drainage basin tributary to the reservoir with dam at point A. Typically, much of the sediment yield from the tributary basin is deposited in the reservoir and the balance is passed through the dam. Zone 2 has been divided into two reaches. In zone 2(a) relatively clear water released from the dam picks up a new equilibrium sediment load over some distance below the dam. The extent of degradation varies with a number of factors among the most important of which are magnitude of dam releases and size and gradation of bed and bank materials. Zone 2(b) is comparable to Schumm's transfer zone, and zone 3 is the area of sediment deposition.

Importance of Sediment Problems in Water Resource Planning

Sediment problems associated with reservoirs include: (1) sediment deposition in the reservoir, (2) distribution of sediment deposits in the reservoir, (3) aggradation upstream from the reservoir, (4) reservoir trap efficiency, (5) reservoir sediment surveys, (6) removal of deposits from the reservoir, (7) degradation downstream from the dam, and (8) sediment abrasion of hydraulic machinery.

All of the adverse effects associated with sediment result in increased project costs. Some factors, such as providing additional storage volume to allow for sediment deposition over the project life without encroaching on useful reservoir capacity, are reflected in project first costs. Other factors cannot be completely foreseen or occur over time, such as aggradation upstream from the reservoir, degradation downstream from the dam, and abrasion damage to hydraulic machinery; these are reflected in annual operation and maintenance costs. Accordingly, all detrimental effects have an effect on economic analysis of a project.

Probably the most critical problem associated with sediment is depletion of reservoir storage due to deposition in the reservoir. Depletion of storage capacity that occurs more rapidly than projected or is greater than projected is a very serious consideration in estimating project benefits. For example, if conservation storage is significantly decreased over the first 20 years of project operation rather than as projected near the end of the 100-year project life, average annual yield will be decreased and future benefits will be less than projected with the full conservation storage available. Similarly projection of flood control benefits is based on flood control space, and if that space is decreased over time, future benefits will be less than projected in planning studies. Where sediment deposits in the vicinity of the dam, such deposits may clog low level outlets or power plant intakes.

Delta deposits at the head of a reservoir and aggradation upstream from the reservoir will result in a rise in upstream water surface elevations for specific flows. This may raise the ground water table and damage agricultural land, interfere with gravity storm drainage and increase local flood damages, block water intakes and sewer outfall lines, and present problems with bridge clearance if the river is navigable.

Scour downstream from dams may lower the ground water level and drain valuable wetland habitat, cause streambank erosion, undermine bridge piers, roads, and other structures along the river bank, cause failure of levees, and so on.

Abrasion of hydraulic machinery is of special concern for hydroelectric plants. Guide vanes and runners of reaction turbines and control nozzles and seats of impulse turbines sustain significant damage from suspended sediment, and costly annual maintenance is sometimes required.

Sediment Deposition in Reservoirs

Estimating Sediment Inflow Volume

Borland (1971) discussed three procedures for estimating annual sediment inflow volume to a reservoir for planning studies. The first two are applicable to areas of little data; the third procedure is based on detailed field measurements, as follows:

1. Field inspection of the drainage basin to determine soil types, main sediment sources (sheet erosion, gullying, flood erosion, channel erosion); comparison of physical characteristics of study area with those of other similar areas for which sediment yield rate (volume per year) is known; and applying that yield rate to the study area to estimate average sediment inflow volume per year.

2. Determine the annual sediment yield rate for existing reservoirs in the general area using field data from periodic sediment surveys of those reservoirs; apply that yield rate to the study area to estimate average sediment inflow volume per year.

3. Compute the total annual sediment load at the proposed dam site using field measurements to determine sediment transport as a function of discharge and computing sediment discharge in tons per day as a function of daily streamflow and total volume of inflow per year.

Sediment Movement and Deposition

The movement of sediment within a reservoir is governed by current and circulation patterns which, in turn, are determined by the effects of river inflow currents, solar heating of the water, and wind. In rivers carrying a heavy sediment load, the inflow may enter the reservoir as a heavy density current underflow. Solar heating results in thermal stratification of a reservoir and eventually complete mixing and turnover of the reservoir water. Wind generates surface waves that rework and resuspend fine sediments deposited in shallow water.

Solar heating is more important in the mid–latitudes in areas of temperate climate. In the mid–latitudes reservoirs are essentially isothermal in late spring, with water temperature about 4°C throughout the reservoir depth. Surface waters warm in the summer due to solar radiation, and the density of near–surface waters decreases as the temperature increases. Finally an upper epilimnion of light warm water rests above a deeper zone of heavier cold water in the hypolimnion. Between these two zones is the thermocline with a high density gradient, as shown in Figure 9.5.

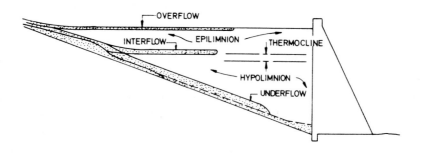

Fig. 9.5 Turbidity Currents and Reservoir Thermal Stratification

In the fall when air temperatures drop, the temperature of water in the epilimnion also drops, and the thermocline disappears. As air temperatures continue to lower, the surface waters eventually become colder than water in the hypolimnium, and the cold surface water sinks below the hypolimnion, resulting in "turnover" of water in the reservoir. This thermal pattern continues throughout the winter until surface waters warm again in the spring, and the process repeats. Tropical reservoirs typically are permanently stratified.

Where river inflows in the summer are warmer and of lesser density than in the epilimnion, the inflow may flow across the reservoir near the surface, as shown in the foregoing figure. If the inflow temperature is between that of the hypolimnion and the epilimnion, the inflow may follow a path along the thermocline as an interflow. If the incoming flow is colder than the epilimnion or carries a heavy sediment load and has a high density, it may flow along the bottom of the reservoir as an underflow.

The pattern of sediment deposition depends on the size and texture of the inflowing sediment, size and shape of the reservoir, the inflow-outflow relationship, and how the reservoir is operated. The coarsest materials (sands and gravels) deposit in the backwater area above the reservoir and in the headwaters of the reservoir, building up delta deposits. The finer silts and clays are carried downstream by density currents, in some cases as far as the dam, and deposit on the reservoir floor or are discharged through the dam.

Backwater deposits raise the streambed upstream from the reservoir. Such deposits tend to be eroded, and some material moves down into the pool when the reservoir operates at a low pool level. Backwater deposits raise water surface profiles upstream from the reservoir.

The U.S. Bureau of Reclamation (1977) found that in most reservoirs the topset slope of reservoir delta deposits closely approximates one-half the original channel slope in the delta area. The pivot point elevation at which the slope changes abruptly approximates the water surface elevation at which the reservoir operates for a large percent of time. The average foreset slope observed in Bureau of Reclamation reservoirs is 6.5 times the topset slope, although some reservoirs have foreset slopes considerably steeper than this. In computing deposition, the volume of sediment deposited in a delta should agree with the volume of material of sand sizes and larger transported by the inflow in the time period considered, assuming a dry weight of about 1200 kg/m^3.

The silts and clays are carried farther downstream into the reservoir and deposited along the bottom of the reservoir and in the vicinity of the dam. The location of these deposits depends primarily on the shape of the reservoir, the mineral characteristics of the clays, and the water chemistry. Where flocculation occurs, the clays are deposited in or near the upstream reach of the pool.

Estimating Reservoir Sediment Deposition

The rate of sediment deposition depends primarily on: capacity/inflow ratio of the reservoir; sediment content of the inflow; and trap efficiency of the reservoir. The rate also depends on characteristics of the inflowing sediment and reservoir operating procedures.

Trap efficiency is the ratio of sediment deposited in a reservoir to total sediment entering the reservoir and depends on the ratio of reservoir volume to inflow (see Fig. 9.6).

Basic steps in estimating reservoir sediment deposition for planning studies are as follows:

1. Estimate sediment inflow to reservoir for specific time increments, for example, yearly.

2. Determine trap efficiency of the reservoir for successive time periods.

3. Determine specific weight of deposited sediment, noting it will change with time due to compaction and with reservoir operating procedures.

4. Project distribution of sediment within the reservoir if distribution is important in planning studies.

5. Estimate loss of reservoir capacity as a function of time throughout the project life.

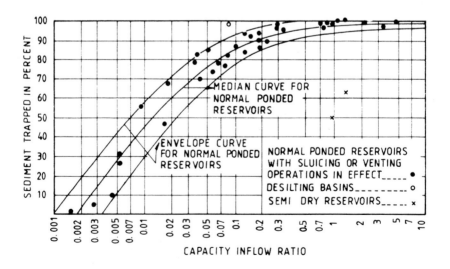

Fig. 9.6 Trap efficiency, after Brune, 1953.

Brune's (1953) curve indicates that virtually all incoming sediment will be deposited in large reservoirs, but his method gives less reliable results for smaller ratios of capacity/inflow where site specific conditions relating to topography, hydrology, and sediment characteristics are of greater importance. The range in flows especially flood rates is a major factor.

Empirical methods such as those of Brune (1953) and Churchill (1948) are adequate for estimates of sediment deposition over time and estimates of reservoir life for planning studies, and the empirical method of Borland and Miller (1958) modified by Lara (1962) is adequate for preliminary estimates of the spatial distribution of deposition. For project design, more detailed studies are needed, and periodic sedimentation surveys are required for project operating decisions.

Sediment Deposition Surveys in Reservoirs

A sedimentation investigation program is an integral part of the overall program for operation of a dam and reservoir. The sediment program is based on periodic resurveys of the reservoir to determine the reduction in storage capacity over time, the distribution of deposits throughout the reservoir, and so on. Resurveys include field measurements, office studies, and laboratory analysis of sediment samples. Field data can be analyzed to determine specific weights of the deposited materials, grain size distribution of the deposits, sediment yield rate of the drainage area, reservoir trap efficiency, density currents, and so forth – all information that is vital to operation of the reservoir and useful, as well, for the design of future reservoirs.

How frequently reservoirs are resurveyed depends on the estimated rate of deposition and on how critical the need is for data on change in reservoir capacity. Resurveys are usually scheduled for intervals of from 5 to 10 years. Sometimes partial or special resurveys are made after major floods.

If sediment inflow volume and deposition are estimated to be large in proportion to storage capacities for various project purposes, accurate records of actual storage depletions are needed for forecasting future depletions far enough in advance to plan and construct replacement facilities. Such information is needed for decisions regarding reallocation of remaining reservoir storage space to the various purposes and the corresponding changes in pool elevations; establishment of elevation limitations and other criteria to regulate construction of boat docks and recreation facilities; revision of reservoir regulation plans to assure optimum utilization of remaining reservoir storage; possible modification of the regulating outlets, water supply intakes, and similar facilities adversely affected by sediment deposits. Accurate data concerning backwater effects upstream from the reservoir are needed if there are problems leading to legal claims upstream from the reservoir arising from operation.

The location of reservoir sediment deposits can be identified by using either contour or range data or a combination of the two. The range method usually requires less time and is less costly; it is widely used. Permanent ranges are monumented in the field and are resurveyed from time to time to obtain profile data that are used to compute changes in volume of sediment deposits.

An accurate reservoir contour map for conditions prior to closure of the dam serves as the basis for estimating sediment deposition by comparison with contour maps based on future topographic surveys; as a basis for determining initial cross section profiles of ranges added to the range network after the beginning of impoundment; and as a basis for determining length factors to be used in computing volume of deposition.

A sediment range is a fixed line across a reservoir along which elevations are determined initially and redetermined in the future, as required, to measure the depth of sediment accumulations or other changes in elevation. The exact locations of ranges should be identified by permanent monuments and vertical and horizontal control surveys.

Ranges are usually located normal to the stream and the valley. They are spaced so that volume, computed by the average end-area method using the cross-sections of adjacent ranges, reasonably represents the volume between the ranges. The spacing of ranges varies with the degree of accuracy desired in the volume estimate. Ranges should be located across the mouths of tributaries and should extend up the major tributaries. A typical layout and a systematic numbering system to identify ranges in large reservoirs uses pairs of beacons at 200–1000m spacing.

Sediment ranges are also established downstream from the dam. In stable channels, characterized by erosion-resistant rock beds and banks, only a few ranges over a short distance below the dam are needed to verify that degradation is not a problem. For alluvial channels, ranges should be closely spaced near the dam and should extend downstream to a point where measurable degradation is not expected to occur in the first 15 to 20 years of operation. Location of ranges below dams on alluvial rivers is influenced by the location of outlet channels, tributaries, and the location of erosion resistant controls.

Field measurements generally include:

1. Survey of established sediment ranges; preparation of topographic maps of special problem areas, etc., to determine elevations and depths of sediment deposits.

2. Measurements needed to compute sediment densities, and sampling required to determine characteristics of the deposited material.

3. Observations, probings, and other measurements not related to established ranges, such as photographs, data on delta areas, etc.

Laboratory analyses are limited largely to analyses of samples of deposited materials to determine size gradation and other pertinent characteristics.

Sediment Management Measures

The problem of sediment deposition in reservoirs can be addressed in a number of ways, some more effective and some more economical than others:

1. Consider the sediment yield of the drainage basin and potential deposition problems in selecting a reservoir site; select a site where sediment inflow will be relatively small. If the reservoir is small and topography permits, select an off-channel site if the river carries a heavy sediment load.

2. Provide excess storage capacity in the reservoir for the sediment accumulation estimated over the project life.

3. Implement watershed management measures to reduce sediment production on the tributary basin.

4. Bypass heavily laden flood flows around the reservoir.

5. Construct debris dams to trap sediment upstream from major storage reservoirs.

6. Provide facilities such as low-level sluices to discharge some sediment through the reservoir.

7. Use mechanical means such as dredging and siphoning to remove deposits from the reservoir.

Reducing Sediment Inflow

The primary means to reduce sediment inflow to a reservoir are: (1) improved watershed management measures; (2) bypassing heavily-laden flood flows around the reservoir; and (3) construction of upstream debris dams to trap sediment before it reaches the main storage dam.

1. Improved Watershed Management Measures to reduce sheet and rill erosion include appropriate agricultural methods, strip planting, terracing, crop rotation, and reafforestation. If the drainage basin is small (1-5 km^2), such measures can reduce sediment yield by 90 to 95 percent. However, for large drainage basins, with numerous landowners, it is usually not physically or economically feasible to reduce basin yield significantly by such methods. At Sanmenxia Reservoir on the Yellow River in China, for example, watershed management techniques were expected to significantly reduce sediment yield, but they have been relatively ineffective.

2. Bypassing Heavily Laden Flood Flows around a reservoir is effective in reducing reservoir sediment inflow, particularly in arid and semi-arid areas. This was done at the Hushan Reservoir (irrigation water supply) in China where in seven years about 54 percent of the storage was lost due to sediment deposition. A small diversion dam was constructed at the head of the reservoir, and flood flows were diverted around the reservoir, reducing the annual rate of sedimentation to about eight percent of what it was originally (UNESCO, 1985). The UNESCO publication also reports that bypasses have been used in the USSR and in Switzerland. However, such plans are expensive and, depending on the topography, not always feasible. It is most often economically feasible for small impoundments for hydropower where bypassing sediment also reduces maintenance costs associated with abrasion of the hydraulic machinery.

3. Debris Dams and Sedimentation Basins have been constructed to trap

and permanently store sediment that otherwise would enter downstream reservoirs. In the U.S., both the Corps of Engineers and the Soil Conservation Service have constructed debris basins in mountainous areas. They are essentially small reservoirs located in canyons in mountain foothills and are designed to trap coarse sediments. Some basins are maintained by periodically removing the sediment by mechanical means. They are an effective control measure.

Future Trends

There is increasing confidence in and use of mathematical models for simulating runoff (ASCA, 1982). Modelling the erosion and deposition of sediment is not as easy and empirical models such as the Universal soil loss equation (ASCE, 1975) are being replaced by physically based models where water runoff velocities are accounted for (Stephenson and Meadows, 1986). The lack of data, particularly rainfall intensities, makes continous modelling difficult and interpolation procedures are generally necessary (Paling et al., 1989). Water resources models based on monthly rain (e.g. Pitman, 1973) have not proved of use for sediment modelling. In any case physically based erosion models (Yalin, 1963) coupled with hydrualic models appear most promising.

REFERENCES

American Society of Agricultural Engineers, 1982. Hydrologic Modeling of Small watersheds.
American Society of Civil Engineers, 1975. Sedimentation Engineering. Manual on Engineering Practice, 54.
Borland, W.M., 1971. Reservoir Sedimentation, in River Mechanics, H.W. Shen, ed. Water Resources Publications. Colorado.
Borland, W.M. and Miller, S.P. 1950. Distribution of sediment in large reservoirs, ASCE Proceedings. Vol. 84, HY2.
Brune, G.M., 1953. Trap efficiency of reservoirs. Trans Am. Geophys. Union. 34 (3).
Churchill, M.A., 1948. Analysis and use of reservoir sedimentation data. Proc. Fed. Int. Sed. Conf. USBR, Denver.
Lara, J.N.M., 1962. Revision of procedures to compute sediment distribution in large reservoirs. U.S.B.R.
Paling, W.A.J., Stephenson, D. and James, C.S. (1989). Modular rainfall runoff and erosion modelling. Water Systems Research Group, University of the Witwatersrand, Johannesburg.
Petts, G.E., 1984. Impounded Rivers, John Wiley and Sons,.
Pitman, W.V., 1973. A mathematical model for generating montly river flows from meteorological data in S.A. Hydrol. Research Unit, University of the Witwatersrand.
Schumm, S.A., 1977. The Fluvial System, John Wiley and Sons.
Stephenson, D. and Meadows, M.E., 1986. Kinematic Hydrology and Modelling, Elsevier, Amsterdam.

U.S. Bureau of Reclamation, 1977. Design of Small Dams.

UNESCO, 1985. Methods of computing sedimentation in Lakes and Reservoirs, Stevan Bruk, Rapporteur.

Vanoni, V., 1946. Transport of Suspended sediment by water. Trans. Am. Soc. Civ. Engrs. III.

Venn, A., 1988. Notes on soil conservation. Continuing Engineering Education Course on Water Resources in Developing Countries, University of the Witwatersrand.

Walling, D.E., 1984. The sediment yields of African rivers, in Challenges in African Hydrology and Water Resources, IAHS Publication No. 144.

Yalin, T.S., 1963. An expression of bed load transportation. Proc. ASCE., J. Hydraulics Div., 89 (H 73). 221 – 250.

Zachar, D., 1982. Soil Erosion, Elsevier, Amsterdam, 547p.

CHAPTER 10

IRRIGATION

LESSONS FROM THE PAST

Many civilizations developed on the basis of irrigation. Egypt, Mesopotamia and China are classical examples and more recently (sixteenth to nineteenth century) in North America and India. There is still similar opportunity for initiating collective action and development in many less developed countries.

Irrigation has long been a prime investment target for development money and foreign aid. It appears an obvious type of investment which would employ local people, generate local wealth and improve health and nutritional standards. A number of books and studies have been orientated towards irrigation projects, and tens of billions of dollars have been spent on such schemes in the past 30 years. A large proportion of the investments by the World Bank and funding agencies in various countries has been in this type of project.

Many early schemes however, were planned using first world criteria or by people who had little understanding of local customs, values and procedures or ambitions in less developed areas. Many projects have become economic burdens in fact as well as environmental and social embarrassments. Land has been eroded, reservoirs have been filled with sediment and canals have been blocked by this sediment or rocks or weeds due to either water or wind erosion. Recently collected data indicates that the financial return of some schemes is only of the order of 15 – 30% of those predicted (Pearce, 1987).

Many of the shortcoming can be attributed to inappropriate engineering. Another reason could be that economists based their projections on incorrect standards and perspectives. They have also not considered the dumping of large quantities of foods from first world countries and the shadow value of local production in indicating poor returns. Some blame has been put on the scale of projects built. That is, vast amounts of money have been put into the construction of dams and waterways, and such investment is out of proportion to funding for associated training schemes and supervision and provision of infrastructure to enable the schemes to be viable and long term. Training needs extend beyond irrigation methods to include conservation and economy as well as marketing and modification of social habits.

In early projects the aspect of water-borne diseases, such as bilharzia and malaria, and effects of the change in flow in rivers due to the construction of dams (and therefore the change in sedimentation patterns in river valleys such as the Nile) probably were not adequately considered.

Irrigated farming makes up about 15 percent of the world's farm land but provides 40% of farm output. While 75 percent of the irrigated area is in the third world, it does not provide enough food for the third world.

Economists such as Smith from the Institute of Development Policy and Management, Manchester University, consider that methods of economic verification are often insufficient. Large scale benefits can be derived using American type benefit cost analysis, and in fact projects can be written off in as short as twenty years on this basis. However the rate of development in third world countries is often very slow and therefore it may take many years to get into gear and get the system working viably, by which time maintenance has probably been neglected.

The relocation of people and villages to accommodate irrigation schemes and provide labour have often been the cause of environmental degradation. Other factors not considered have apparently been the suitability of soils, the long term effect of irrigation on the soil and change in water table level. In fact the amount of research spent on irrigation is tiny compared with other industries. The annual yield of irrigation has been estimated at 100 billion dollars making it 30% of the size of the oil business, but the proportion spent on research was probably in the ratio of 1 to 20 i.e. research for irrigation and associated disciplines was inadequate. A large cause may be that such irrigation schemes are out of sight of the average person and of students who, therefore, do not clamour to get into that field of research whereas they are well aware of oil shortages and the vast sums of money flowing through oil companies. Associated with soil and crop type studies should be hydrological networks, reliability and irrigation scheduling. Social and economic studies are almost lacking on such projects as they are regarded as engineering projects; i.e. inter-disciplinary type work is required.

Planning of major irrigation schemes has generally been of a high standard with respect to hydrological and engineering but, again generalising, of poor standard with regard to soils, enterprise options and agricultural economics. Furthermore, the evaluation of plans in the past has been inadequate, usually featuring only technical and financial evaluations and ignoring social and economic costs and benefits.

Irrigation is a facet of a dynamic system that is in a constant state of change. In the national interest this needs to be monitored. Monitoring is conspicuous by its absence in some important respects e.g. (typically) land use, soil status, micro and macro economics, catchment character and "health", and poor in some others e.g. (typically) water quality and private sector abstractions. This has resulted in a series of crises in the past. More will follow in the future.

Conveyance systems rarely contribute fully to the beneficial use of water for irrigation, mainly for lack of innovative thinking stemming from a lack of multi-disciplinary planning. Conveyance systems can be long, and they may traverse soils of poor quality, but usually can be made to contribute materially to the viability of irrigation projects.

Water for irrigation is, properly, a third priority issue and there is general agreement on the need to improve the efficiency of water use in irrigation, yet little is being done to bring this about. Many users are unaware of even the concepts of such important basic issues as water scheduling.

Large-scale formal irrigation schemes in developing areas have, for the greater part, been a failure and, engineers must bear responsibility for many (but not all) of such failures.

One of the lessons of the past is that engineers should place greater emphasis on the non-engineering skills needed in water use, research, planning and monitoring, and should become more sophisticated in their evaluation of proposals for the use of water for irrigation.

The most productive and the most efficient production unit in the underdeveloped rural areas, not only in terms of agriculture but also in terms of other economic activities, it the homestead site, which is commonly enclosed. This efficiency stems from such factors as the (a) natural and substantial addition of fertility to the homestead, (b) availability of labour there, (c) local demand for products by the family and neighbours (assured doorstep market) and (d) convenience and the more efficient use of time. (Witness gross agricultural product value, greater Maseru, $5 million p.a.)

Given a supply of water for irrigation and other productive purposes, the homestead can become an even more efficient producer of food and other goods, without the need for many of the costly features of orthodox irrigation schemes (complex systems, external management, farmer servicing, etc).

The homestead site has other very important features. It is the best conserved part of the countryside, and the main locality for natural capital formation, as opposed to the resource rape that takes place elsewhere. Even in the absence of formal title, tenure is more secure there than on any other part of countryside where the householder may have rights. Other favourable features include its being the place where there may be light in darkness, the place of learning, the place of social intercourse, and so on.

Under these circumstances it makes sense, where feasible, (under the adverse circumstances of the elevated location of most villages) to provide water for micro-agriculture and other productive purposes at the homestead site. The main disadvantage of this approach is the relatively high capital cost of infrastructure but this is largely a "one of" cost, especially where, as recommended, the source of water is essentially local and the water supply system is essentially parochial and community based, rather than a regional "big scheme" system.

Homestead micro-agriculture introduces the need for new approaches in town planning – particularly in (a) the compromise between plot size and the costs of water reticulation systems and other infrastructure, and (b) the optimum location of new towns and villages or the expansion thereof.

Homesteads micro-agriculture also introduces the need for innovative new approaches in water engineering particularly with regard to conveyance, storage and metering systems in villages. Homestead micro-agriculture has begun to develop. It represents but one system of using water for irrigation but it has an important place in the future planning of optimisation of use of water.

SMALL SCALE IRRIGATION SYSTEMS

The International Fund for Agricultural Development (IFAD) has recognized that large scale irrigation may not be the answer in developing countries and is investigating small scale type projects which bring basics to the local people and result in less of a disruption than the large scale type engineering project.

It is recognized that lack of water is a major constraint in many developing countries, in particular the Sub Saharan countries in Africa. In fact, small scale irrigation schemes are probably the only viable ones in that area. For this reason is not widely practiced in much of Africa, irrigation and the small scale type would perhaps be more acceptable than large scale projects.

In some cases irrigation is merely supplementary and in some cases rain fed agriculture would provide a limited crop i.e. problems with the dam or conduits may not be a disaster. In other cases no form of agriculture would be possible without irrigation, in and a complete new infrastructure and way of life must be introduced with the irrigation schemes. In some areas irrigation has taken off and is now self propagating. For instance in Nigeria half a million hectares of ravine valleys have been developed in the past 20 years by local farming communities with little outside help. The rate of growth is 10 percent per annum.

The cropping needs careful consideration. Initially it may be wise to grow subsistance type crops to make people aware of the value of the irrigation, but as soon as possible introduction of patch crops and export crops not used by the local people would improve the economy.

In the drier areas methodology for managing with little water needs further development. In countries such as Israel drip irrigation and fine mist sprays have reduced water consumption considerably. The same applies to equipment, and threshing and reaping equipment obviously needs to be designed to suit the specific project. For instance, manually operated machinery is often preferred to large mechanical equipment which both breaks down and does not use local labour.

IRRIGATION TECHNOLOGY

Procedures for designing dams, canals and diversion works have been established through colonial experience in India and Egypt for example (see Houck, 1952), but these are generally for large scale works.

Modern irrigation systems are considerably more efficient than the older methods. Traditionally irrigation has been by flood in developing countries. Sprinklers however are reputed to be much more water efficient although they are more capital intensive. Another technological advancement is the automatic type of lateral valves and schedulers. They again have proved efficient in advanced countries but may not be understood or may be mismanaged in less appropriate environments. Travelling irrigation sprinklers and even centre pivot type schemes require careful operator training before they are viable. In general, the use of the smaller drip type nozzles appears to be less technological but requires more maintenance in the way of cleaning of nozzles etc. Computer crop scheduling is also obviously not appropriate unless permanent contract type managers are brought in to assist with the operation i.e. a back up economic input is required for many years before such projects can be taken over by the local people in many cases.

Methods of distributing water have been developed in third world countries which are most appropriate to the circumstances. For instance the surjan which was developed in Java (Pickford, 1987).

The surjan is a system of parallel furrows through which the water is diverted and more particularly rainwater is caught and held. Different crops are planted in the furrows and on the ridges. For instance trees, which have deeper roots, may be planted on the ridges and even rice could be planted in the furrows.

The Food and Agricultural Organization of the United Nations (Pickford, 1987) has started a review of existing irrigation schemes. They are concentrating on small scale methods such as the construction of wells. They are also looking at resettlement and provision of roads and infrastructure to ensure the projects are viable. They consider the necessity for continuing with livestock despite the problems due to poor grazing practices, realizing that keeping livestock is a basic part of life in many situations either acting as wealth indices or means of transport. Often the use of livestock for food is of small consideration as it is expensive to rear when considering the amount of grazing land required.

There are also many mission stations which are improving agricultural practices. By forming a focus, the awareness of standards of living and appreciation of the values in life are instilled in the people. A sense of permanence and, therefore, awareness of the environment and improvement in living standards results. This in turn leads to a desire to improve crops and therefore irrigate and fertilize and manage well.

Considering the number of people in the world involved in agriculture and irrigation, and in fact only in these fields, then the amount of money spent on research and even understanding their problems is small and probably at the moment insufficient to hope to raise the living standards of the majority of the world's population.

In underdeveloped areas the competition for water is not likely to be fierce. It may be more appropriate to design a low efficiency system with resulting savings in costs. Unlined canals around flood irrigation may be most appropriate for small scale plots. Where the plots are scattered saturation of the subsoil may not be a problem.

The net result is that more area and hence more people, can be supplied for a fixed budget. Operating manpower is likely to be higher but less skilled than for sprinklers or drips. That is, however, one of the objectives i.e. to provide employment for rural people and reduce use of pumps, pipework and complex application systems to a minimum.

PLANNING IRRIGATION

Land classification is an important but often neglected aspect in planning irrigation. Soil composition is important from crop growth point of view, texture for drainage and moisture retention, salinity for affecting the transpiration process, slope for drainage and access, and depth for ploughing and drainage. Stone content, weather, ease of initial cleaning, aspect for sun and elevation for supplying water require further consideration. Then ownership, previous use, social attitudes and present cover will influence the decisions as well as economics of water supply, land preparation, fertilizing, crop suitability, weeding, reaping and marketing. Finally human resources, training and management must be available.

Salts in the soil and water can affect productivity. Alkalis can be brought to the surface by capillary action and cake as water evaporates. They also affect the transpiration process and yield. In such soils overhead sprays may be preferable to flood irrigation.

Losses due to seepage and soil evaporation can be greater with flooding and unlined ditches, and the water table can rise causing saturation and flooding.

WATER REQUIREMENTS

The amount of irrigation water depends on:
Method of irrigation
Rainfall
Temperature
Wind
Humidity
Groundwater
Effectiveness of water
Increased yield
Salinity
Crop
Crop density and foliage
Number of crops per year
Time of year
Losses in conveyance, storage and soil

214

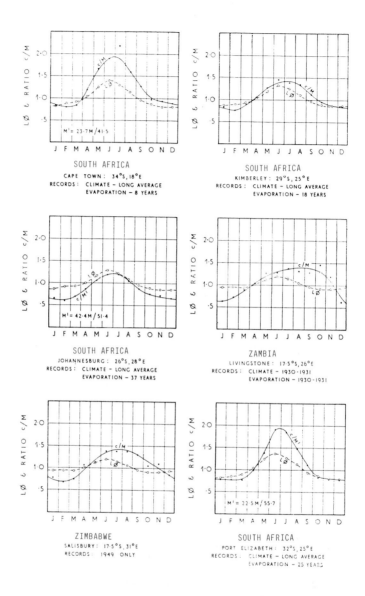

Fig. 10.1   Southern African : Comparison of monthly c/M ratios
with LØ ratios for different Latitudes (Olivier, 1961)

Generally the amount of water is measured as an equivalent depth over the area to be irrigated.

Rate of application can be 20 to 40 mm per day, depending on the soil type and method of application, but the period between applications will depend on crop requirements. As a rule of thumb, 0,5 to 1m of water is required per year, including field losses which can be between 10 and 50 per cent. Transmission losses can be equally large.

Crop requirements can be calculated using a formula such as that of Olivier (1961) based on a modified Penman approach (1948):

$$CuF = Mp/L$$

where CuF is crop use in mm/day at latitude F, L is R/Rv the ratio of actual to vertical radiation (Rv = R sin h where h is the angle of inclination of the sun at that latitude.) c/M is calculated to measured evaporation and Lϕ is L/Lo, the variation of actual to vertical radiation (see Fig. 10.1).

Mp is the evaporation in mm/day from a tank at the same latitude.

An alternative formula is that of the U.S. Department of Agriculture (1974):

Daily potential evapotranspiration in mm

$$E_{tp} = 0.000673 \times 25.4 \ [C_1 \ (Rn-G) + 15.36C_2 \ (1.1+0.017\times0.625W)(e_s-e_d)]$$

where

C_i and C_2 = mean air temperature weighing factors ($C_1 + C_2 = 1$)

e_s = mean saturation vapour pressure in mb

e_d = saturation vapour pressure at mean dewpoint temperature

W = total daily wind movement, km

R_n = daily net radiation in cal/cm^2

G = daily soil heat flux in cal/cm^2

c_2 = $0.959 - 0.0125\bar{T}+ 0.000045 \ \bar{T}^2$

$\bar{T}$ = mean daily air temp in °F

= °C $\times$ 1.8 + 32

$e(T)$ = $-0.7 + 0.295T - 0.0052T^2 + 89 \times 10^{-6}T^3$

G = $5[\bar{T} - (\bar{T}_{-1} + \bar{T}_{-2} + \bar{T}_{-3})/3]$

$\bar{T}_{-i}$ is mean air temp for ith previous day

R_n = $0.77R_s[\dfrac{0.9Rs}{R_{bo}} + 0.1] \ R_{bo}$

R_{so} = solar radiation on a clear day

= $760 \ exp \ [\dfrac{Day1 - 107}{157}]^2$

where Day 1 = March 1 in Northern Hemisphere, or September 1 in Southern Hemisphere.

216

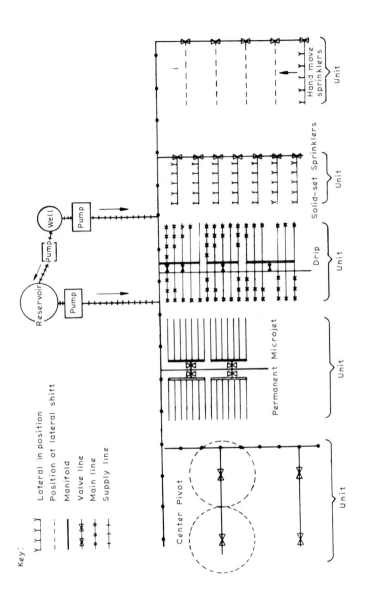

Fig. 10.2 Types of distribution (Karmeli et al., 1985)

R_{bo} = net outgoing longwave radiation on a clear day

$$= (0.37 - 0.044 \ /e_d\ )(11.7 \times 10^{-6})[\frac{T_a{}^4 + T_b{}^4}{2}]$$

T_a and T_b = max and min daily temperature in °K.

As a rough approximation the potential evapotranspiration can equal reservoir evaporation, which is some 20% less than 'A pan' evaporation. With dormant periods the evapotranspiration can be considerably less than free surface evaporation; however field and conveyance losses can add up to 100 percent on to net requirements.

SELECTION OF EMITTERS (see Fig. 10.2)

Factors and Objectives in the Selection of Emitters

The selection of a given type of emitter (drip, sprayer or sprinkler) for low intensity irrigation, is based on a number of factors as follows:-

The nominal emmiter discharge

The nominal emitter operating pressure

The relationships of emitter discharge and pressure

The size of flow cross section (nozzle size in sprayers and in sprinklers; orifice or flow path size in drippers).

The vertical angle of water jet (for sprayers and sprinklers).

The wetting diameter of a single emitter

The wetting pattern of a single and/or a group of emitters

The spacing and position of emitters along and between laterals

The selection of the emitter based on the determination of the factors listed above, is carried out by simultaneously satisfying a set of objectives which are directly affected by the emitter characteristics. The basic parameters in each case:-

i) The application rate of the irrigation.

$$I = \frac{qE}{b \times r} \times 1\ 000$$

Where   I = application rate   – mm/hr

qE = nominal emitter discharge – m^3/hr

b,r = emitter spacings   – m × m

AGD = gross application depth   – mm

ii) Time of application

The time required for the desired depth of application is given by:-

tapp = AGD/1

Figs 10.3-4 Show the effect of different application rates on yields.

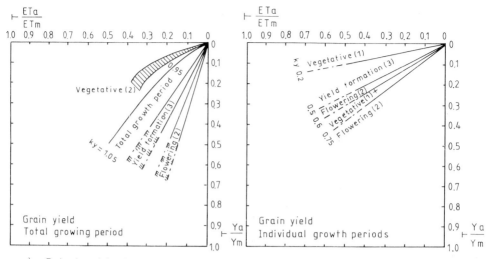

a) Relationship between relative yield decrease (1 – Ya/Ym) and
    relative evapotranspiration deficit (1 – ETa/ETm) for winter wheat.
    (Doorenbos and Kassam, 1979)

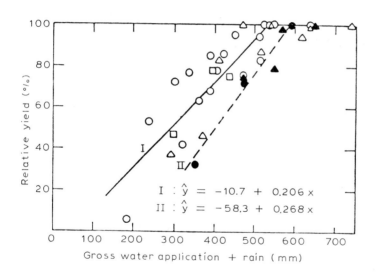

$$I : \hat{y} = -10.7 + 0.206\,x$$
$$II : \hat{y} = -58.3 + 0.268\,x$$

b) Relation between relative wheat   Bet Shean Valley ▲
    yield and total (rain + irrigation)   Jordan Rift   ●   ----
    water application.   Lakhish Region ˙ △
    Shalhevet et. al. (1976)   Northern Negev   ○□   ——————

Fig. 10.3 Crop yield versus water application for wheat.

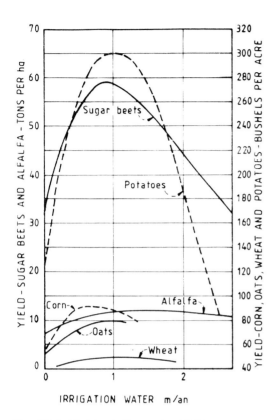

Fig 10.4 Yield of Crops in Utah

REFERENCES

Doorenbos, J. and Kassam, A.H., 1979. Yield Response to Water. FAO Irrigation and Drainage Paper No. 33, Rome 1979.

Houck, I., 1952. Irrigation, ch 17. In Davis, C.V., Handbook of Applied Hydraulics. McGraw Hill.

Karmeli, D., Peri, G. and Todes, M., 1985. Topics in Irrigation Systems Design and Operation – Course in Continuing Eng. Educn. University of the Witwatersrand.

Olivier, H., 1961. Irrigation and Climate. Edward Arnold, London.

Pearce, F., 1987. A watershed for the third world irrigation. New Scientists, p 26-7.

Penman, H.L., 1948. Natural evaporation from open water bare soil and grass. Proc. Royal Soc. No. 1032, 193.

Pickford, J. (Ed.) 1987. Developing World Water. Grosman Press, p 260-309.

Shalhevet, J., Montell, A., Bielorai, H. and Shirushi, D., 1976. Irrigation of field and orchage crops under semi-arid conditions IIIC, Public. 1, Volcani Cr, Bet Dagan, Israel.

U.S. Dept. of Agriculture, 1974. Scheduling Irrigations Using a Programmable Calculator. Publicn. ARS-NC-12.

CHAPTER 11

RURAL WATER SUPPLIES

INTRODUCTION

The concept of community participation in community development started on a small scale in the 1950's. The concept then broadened and grew during the 1970's, stimulated, in part, by the failure of many of the large development projects of the previous decade; today it has become a significant factor in development planning. The early models of development, in the immediate post colonial era, were purely economic and were based upon the "trickle down" hypothesis, whereby (in simplistic terms) economic growth or benefits in one sector (e.g. urban industrial development) would eventually spread to other sectors such as the rural economy. This model has proved to be invalid and, if anything, the reverse has happened, in that, while a small elite has prospered, the large majority of the rural population has become poorer.

There are exceptions to the above of course, and undoubtedly there have been cases where large scale investment in specific geographical areas (generally those with good agricultural potential) have benefitted the majority in the area. Thailand is one such case. However, the Thai experience also showed that there was no spread of benefit outside the particular geographical area being assisted and that the net result was, in fact, an overall increase in the gap between rich and poor nationwide.

TECHNICAL ASPECTS

Surveys carried out by the World Bank indicate failure rates in water and sanitation projects in developing countries as high as 80%, while surveys of boreholes in many areas reveal that as few as 35 percent of existing boreholes are operational at any one time.

This observation in no way diminishes the importance of technical choice in a rural water supply project since, ultimately, the scheme must be technically viable. The chart in Figure 11.1 shows a range of rural water supply options most likely to be encountered. Basically the system can be divided into three parts : source; treatment; and reticulation.

Table 11.1 highlights three sources of supply : groundwater, surface water and rainwater. Rainwater, while capable of providing a supplementary supply, is unlikely to become the major source due to the

WASHING AND DRAWING WATER
FROM STREAM OR POND

BOREHOLE, WINDMILL
AND CONTAINER

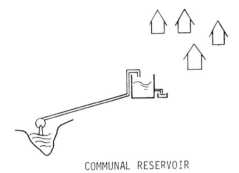

COMMUNAL RESERVOIR

SLOW SAND FILTRATION

TOWNSHIP STANDPIPES

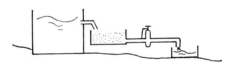

PURIFICATION AND DISINFECTION

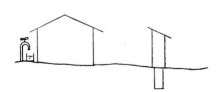

SINGLE DOMESTIC CONNECTIONS

MULTIPLE HOUSE CONNECTIONS,
HOT AND COLD, AND SANITATION

Fig. 11.1   Evolution in Water Supply Standards

limited rainfall available. Surface waters too are limited in potential to specific areas of many countries, although it is expected that usage from this source will increase rapidly in the near future. However, at the present time groundwater is probably the major source of water for the majority of the rural population throughout the world and is likely to remain so (Uphoff, 1979).

Typical rural water consumption figures vary from 10 to 30ℓ per capita per day. Interestingly the figure would appear to be relatively independent of distance travelled to fetch water. Accurate figures have been obtained from studies in Swaziland where there is a major rural water supply programme. Here water consumption for domestic use (use of water for gardening is not permitted from the domestic supply) averages 25ℓ/person/day. Given that water is collected over a 12 hour day on average, and assuming a turn-around time of between 2 and 5 minutes at the pump, then between 20 and 50 families can be served by one handpump. (The Lesotho objective is one pump per 20 families.)

The next stage in the development of the water system is a fuel driven pump, high level storage tank and simple water supply network to standpipes. Here much work still needs to be done in the area of rationalisation, with equipment being standardised throughout a geographical region at least, if not throughout the country. Zimbabwe and Mozambique have reduced handpump types to three and two respectively and have standardised on one type of material for pipe construction, and other countries are beginning to follow that philosophy.

Having eliminated options which are technically unsuitable, the final choice needs to consider not only financial constraints, but also the issues of organisational structure, education and training. An illustration of one approach is given in the following case study of the Lesotho rural water supply programme. This study is interesting for several reasons. Not only is the approach well documented (M.B. Consulting, 1987), but in addition it shows a clear use of a specific community participation model, which corresponds to Oakley and Marsden's Community Development model (1984). A similar system is also operational in Swaziland.

CASE STUDY

The Lesotho system relies strongly on the traditional village management system where there is a chief and a village council responsible for all aspects of development. The relationship between village and central government is handled by a rural development office, whose work is to explain government policy, motivate the community for

TABLE 11.1   Rural Water Supply Options (Abbott, 1988)

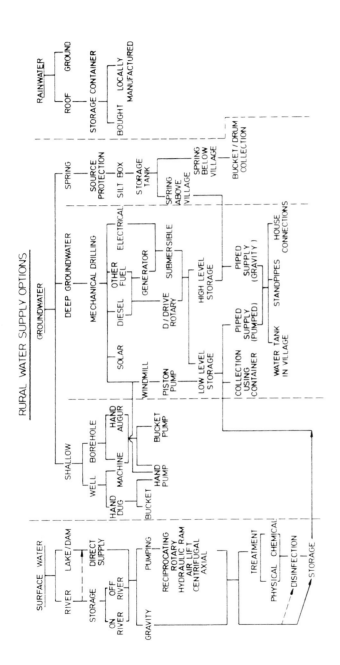

village water supply work and encourage those who seem reluctant to participate. The office also provides administrative assistance in setting procedures, etc. There is also a technical division within Government, called the village water supply section or VWS whose function is to choose the best technical option for a particular village and to work with people to construct the village water supply. Before a water supply can be installed, a water committee which is correctly constituted must first be established. Because Central Government pays for the installation it determines priorities within the country, based upon factors such as need (e.g. no. of people served, whether schools and clinics are present, potential for an outbreak of disease in the area), the degree of community interest, and technical possibilities for construction. However, once the system has been installed, ownership rests entirely with the village.

In terms of construction the VWS first sends in a mason to teach the villagers to shape stones for storage tanks, to dig furrows for pipes and to help in organising the labour force. When two thirds of the work is complete the full VWS team will arrive and supervise the more complex aspects of the project. Once completed, the village is responsible for looking after the system and for carrying out minor repairs. Maintenance and minor repairs would normally be carried out by a water minder, appointed by the village and trained by the VWS, while for supervision of the supply and distribution system a group of villagers called guardians would be appointed. Each committee has a treasurer who is taught simple bookkeeping, and each village then opens a special bank account. All villagers contribute and the income is used for spares, diesel and for paying the water minder. Major repairs are handled by the VWS. The whole question of financial management is seen as crucial to the success of the project and a significant proportion of the training effort is devoted to this issue.

It can be seen that, with this system, central government provides a "safety net" for the village. Nonetheless there is a degree of decision making at a local level which would appear to be satisfactory. Certainly both the Lesotho and the Swaziland schemes appear to be working well at this stage.

DEVELOPMENTS IN SUPPLY

Considerable effort has gone into supplying rural population with potable water (e.g. ECLA, 1973).

Due to natural preferences and relocations, villages often occur along the crests of hills i.e. at the furthest points from rivers and also where the water table is deepest. This results in a large proportion of the time of women and youths (sometimes over 4 hours a day) being taken up in transporting water from rivers and springs to the villages. This tedious method of collecting water has reduced the water consumption to a minimum, namely less than 4 litres per capita per day in many cases.

The World Health Organization has suggested a minimum water supply of 20 litres per capita per day which would appear impossible with many present situations (1982).

Attempts to Supply Water to Rural Population

Villages can be classified depending on the need for water supply. That is, critical areas can be defined as those where there is less than 10 litres per capital of water available per day and access is over a thousand metres to the nearest water source and the ground slope along that way is steeper than 12%. On the other hand those which have been 10 and 20 litres per capita per day of water available and live within 750 to 1000 metres of a water source with a slope of between 6 and 12% are classified as needy and the balance is classified as having adequate water supply.

Cost of Rural Water Supplies

Prior to assessing the overall investment required to provide adequate water, the system of water supply has to be optimized. In other words, the most economical method of supply was considered as well as the method of conveyance and distribution.

The options open for supply of water include (see Table 11.2):

1) Groundwater

There are not many natural springs in drier areas owing to the low permeability of the underlying rock, and the groundwater table is often far below the surface. It is therefore generally necessary to pump, for example with a windmill, as hand pumps can often not cope with the high lift required. It also follows that the average cost of drilling boreholes is relatively, high and (together with pump and pipeworks) the cost per borehole plus windmill can be as high as $60 000 per scheme with reticulation. including a reservoir. The typical yield of such a windmill

TABLE 11.2 Community Water Supply Services in Developing Countries
(Source: World Health Organization, 1973) (Data as at December 31, 1970)

Region and Country	Urban population supplied						Rural population with reasonable access		Total population supplied	
	By house connections		By public standposts		Total urban					
	N'000	%	N'000	%	N'000	%	N'000	%	N'000	%
Summary for all developing countries:										
Africa	8 876	29	11 921	39	20 797	68	16 717	11	37 514	21
Americas	95 410	60	26 724	17	122 134	76	29 549	24	151 683	54
Eastern Mediterranean	38 093	59	16 726	26	54 819	84	31 255	18	86 074	33
European Region	12 406	50	5 426	22	17 832	73	18 400	44	36 232	55
South-East Asia	56 391	36	26 798	17	83 189	53	61 095	9	144 284	17
Western Pacific	25 107	65	3 668	10	28 775	75	16 067	21	44 842	40
Total	236 283	49	91 263	19	327 546	68	173 083	14	500 629	29
Africa										
Botswana	16	46	19	54	35	100	149	25	184	29
Burundi	15	15	60	62	75	77	-	-	75	2
Cameroon	150	13	750	64	900	77	1 000	21	1 900	32
Central Africa Republic	16	4	34	9	50	13	-	-	50	3
Chad	30	11	170	65	200	76	780	22	980	26
Congo	80	28	198	69	278	98	46	7	324	34
Dahomey	33	9	313	86	346	94	455	19	801	29
Gabon	5	5	1	1	6	6	1	n	7	1
Gambia	10	27	26	70	36	97	9	3	45	12
Ghana	652	22	1 483	51	2 135	73	870	14	3 005	33
Guinea	337	75	100	22	437	97	-	-	437	11
Ivory Coast	260	28	656	70	916	97	1 000	29	1 916	44
Kenya	1 000	90	72	7	1 072	97	240	2	1 312	12
Lesotho	5	19	22	81	27	100	-	-	27	3
Liberia	60	43	50	57	140	100	67	6	207	17
Madagascar	236	25	594	63	830	87	45	1	875	12
Mali	160	26	20	3	180	29	-	-	180	33
Mauritania	80	91	6	7	86	98	114	10	200	17
Niger	40	12	180	55	220	68	570	16	790	20
Nigeria	2 810	22	4 650	36	7 460	58	3 586	8	11 046	20
Senegal	300	29	722	69	1 022	98	2 178	74	3 200	81
Sierra Leone	102	27	180	53	282	75	26	1	308	12
Togo	34	13	214	84	248	97	86	5	334	18
Uganda	400	58	216	31	616	89	1 600	20	2 216	25
United Republic of Tanzania	100	11	400	44	500	54	1 200	10	1 700	13
Upper Volta	40	20	100	49	140	68	1 300	25	1 440	25
Zaire	1 205	41	400	14	1 605	55	750	5	2 355	13
Zambia	705	71	255	25	955	97	645	19	1 600	37

could be 15 cubic metres per day which would provide water for 800 people.

It may be noted that there is a severe limitation on hand pumps in such circumstances since they can only deliver about 8 litres per minute, and the number of working hours a day would therefore limit the yield of a borehole to less than 4 000 litres per day. It would also be necessary to collect the water on foot, and the physical effort thus required would place a severe limitation on the system.

2) Surface Water Resources

There are many regional water supply schemes which are operating successfully, but, in some have maintenance problems. It is recognized that such schemes can meet minimum standards in quantity of water available, and generally the water quality can be adequately monitored. The capital cost of such schemes can, however, get out of hand unless it is considered in the perspective of the benefit achieved. For instance a dam, a pumping station, a pipeline and purification works would all normally be associated with such a scheme. The cost per capita and per cubic metre is highly dependent on the scale of the project and for small scale supplies the cost may easily exceed $10 per cubic metre of water delivered. On the other hand, Figure 11.2 indicates that the cost reduces by orders of magnitude if the population supplied increases by the same order. The cost also does not allow for reticulation but assumes a bulk supply with distribution from a central stand pipe or pattern of stand pipes.

3) Rainwater Collection

Limited research has been done in the field of collecting rain by impermeable covers and diverting it into a tank. The reliability and the cost however appears to rule this method out in most cases.

VALUE OF WATER

Injecting money into water supply schemes could, if managed properly, result in improvement in living standards due to the creation of employment and circulation of money. On the other hand, to provide an isolated construction job may be disrupting to the society although it provides short term employment. Development authorities, therefore, appear

228

to favour integrated development i.e., one scheme after another providing continuity. This encourages local responsibility, awareness campaigns and labour intensive methods. The cost and need are not the only criteria in deciding whether to supply a rural village with water. Even with the most economical type of supply throughout, the cost of supplying everyone in the world with water would be over $600 billion. Whether everyone would benefit appreciably from such water supplies and, in fact, whether supply to everyone is justified, perhaps needs reassessment until there is reasonably full employment and the people's time could be put to better use. Perhaps labour intensive type water supplies may in fact be meeting many of the objectives sought in water resources development (provision of employment as a means of circulating money and improving the economy of the country). Labour intensive methods range from hand drawing and transport by means of animals, to simple gravitational feeds with basic purification.

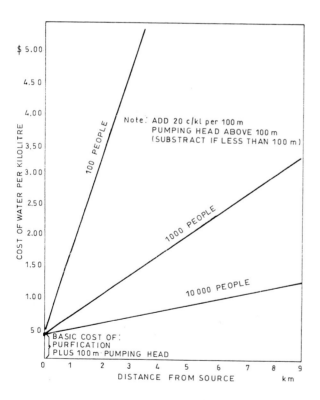

Fig. 11.2 Typical Cost of Rural Water Supplies

An alternative would be to supply water in bulk at selected positions, probably where the cost is reasonable and other services are available. The people, if they require water or desire water, may then relocate to where the services including water are adequate. One could then detect from migration the desirability of having such services amongst the community. Affordability of such services must also be considered. It is realised many communities cannot hope to pay for water, let alone electricity. However, if the economy were to grow to such an extent that there would be reasonable employment and people could be charged a nominal fee or eventually the full fee, this would be a great step towards a free economy. The possibility of coin box pumps could also be considered.

One advantage of centralized regional water supply schemes is that they form a node for settlement. A community, and later a village or town, could be so established, resulting in a net urbanization, which could take some pressure off the land. The continued support of scattered rural huts is having a devastating effect on soil conservation and fertility and is denuding areas of natural vegetation and trees.

OPTIMIZATION OF BULK SUPPLIES

Bearing in mind the advantages of scale in water supply, a methodology for a) ranking water supply schemes and b) optimizing the design of such schemes is being developed. It is implicitly assumed in ranking that the schemes have been initially optimized i.e. that the cost of each alternative has been reduced to a minimum. This must be done by systematic methods.

Optimum Design of Distribution Networks

An example of a water supply scheme for a regional community or scattered villages is given below. The scale, i.e. size of pipes, as indicated previously, has an important bearing on the cost per unit of water supplied, and therefore the routing of the pipes should be such that the total supply carried through each pipe is as high as possible. There are, however, many possible routes and alternative branch type networks to supply villages within a district. Assuming that each village requires a supply of 20 litres per day per capita, the network can be set up as a linear programming exercise. The results of such an analysis are indicated in Figure 11.3, i.e. the optimum supply route and amount of water

230

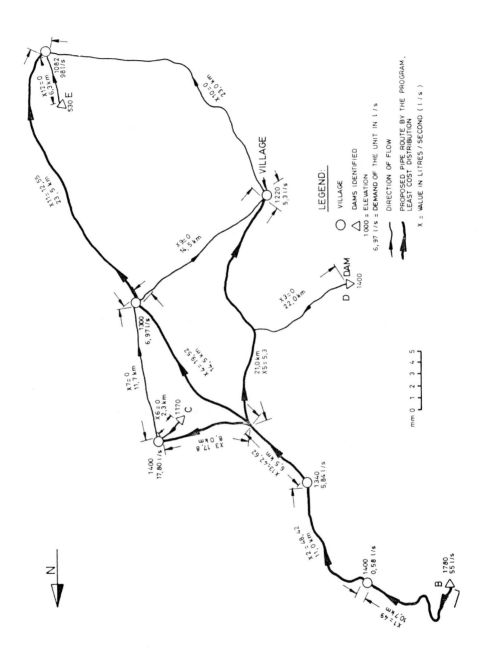

Fig. 11.3 Alternative Water Supply Patterns to Villages

LEGEND:

○ VILLAGE

△ DAMS IDENTIFIED

1 000 = ELEVATION

6, 97 l/s = DEMAND OF THE UNIT IN l/s

DIRECTION OF FLOW

PROPOSED PIPE ROUTE BY THE PROGRAM.
LEAST COST DISTRIBUTION

X = VALUE IN LITRES / SECOND (l/s)

mm 0 1 2 3 4 5

supplied the villages are indicated.

APPLICATION

As a case study, the water supply to Umzimkulu was studied. The area embraces 190 000 people in 133 villages which have at present delivered water and the least cost distribution system was to be selected (Fig. 11.3). The optimum source and bulk supply pattern was selected by computer linear programming techniques (Stephenson, 1984). The resulting pattern proved to be 20% cheaper than the best alternative obtained by hand.

Shadow values were then applied to construction costs for use of labour. The resulting distribution pattern was however not affected in this case as all construction and operating prices were reduced by about the same proportion i.e. 35%, to allow for labour utilization.

WATER SUPPLY INDEX FOR RANKING PROJECTS

In order to attempt to assess and prioritize projects, the variables were set down in a formula as follows.

Let W_1 = water supply rate in kℓ/day, averaged over the first 10 years of operation, for domestic, stock and other uses where the consumer cannot meet the actual cost.

W_2 = water supply rate in kℓ/day to industries and other consumers who can pay the water costs, averaged over the first 10 years of operation.

C_1 = capital cost of proposed project in Dollars, including labour proportion x. Use $(1-x)C_1 + xC_1 K_1 = C_1$ for WSI

C_2 = operating cost \$/annum of pumping, maintenance, treatment with chemicals and labour, allowing for labour component as in C_1

Labour cost = marginal cost only, i.e. K_1 times salary.

where K_1 = 0 for no alternative employment and local funding.

= 0.5 for no alternative employment and foreign funding

= 1.0 for zero unemployment areas i.e. competing employment opportunities

R = availability as a fraction of time (less than 1 due to breakdowns, no supervision etc).

Then the water supply index, $\text{WSI} = \dfrac{100 \ (C_1 \times 0.087 + C_2)}{365R \ (W_1 + W_2)}$

The discount factor 0.087 is based on 6% real discount rate over 20 years.

The index should be optimized (minimized) for each project before ranking, e.g. by deciding on minimum treatment needed, minimum design supply rate per capita and maximum scale of development. Then alternative schemes can be ranked with highest priority for projects with low WSI.

APPLICATION OF WSI

Applied to alternative sources of water for the selected case study, the technique produced the following indices:

Borehole and windmill WSI = 40 - 120 (40 based on 80% availability, 120 based on 40% availability and costly maintenance)

or WSI = 140 - 240 with reticulation

IRWSS WSI = 100 - 140 reduced to -100 with optimization.

Borehole, pump and reticulation WSI = 130 - 280

The use of a water supply index has enabled alternative rural water supply schemes to be ranked. The ranking is not only based on cost, but includes factors for involvement of local people, reliability, capital versus operating intensive cost and phased construction. In general, the integrated regional water supply scheme appears the most attractive solution provided the layout and general scheme plan is optimized carefully, as indicated in the summary in Table 11.2.

TABLE 11.2

Type of scheme	Limit of capacity	Capital cost	Cost c/m^3	Problems
Handpump	4 000 ℓ/d	$5 000	40	Low Yield/ capacity
Windmill + dam	15 -20 000 ℓ/d	$10 000	20	Maintenance
River pump	5 000 000 ℓ/d	$5 000 000	30	Power, scale
Rainharvesting	8-15 000 ℓ/d	$20 000	60	Storage, cost

All costs exclude reticulation and purification. No conveyance pipe was allowed for handpumps and rainharvesting.

WATER QUALITY

It is no use having water supply for drinking and other domestic purposes without paying attention to the water quality as well. Many sources of water can be dangerously polluted, and the investigation into water supply needs to analyze the water and install appropriate purification systems. Where the purification is not likely to work, reliable alternative sources must be sought. The following sources of pollution have to be considered.

Stagnant water: could also be used for washing ablutions etc. and likely to be severely contaminated with biological matter, parasites.

Ground water: could be polluted by pit latrines, human or stock pollution, dangerous nitrates and bacteria.

River water: often silt laden especially during flood; requires settling and filtration. Could be contaminated by upstream industry.

Rain water: Dust, nitrates, sulphates possible, depending on wind and industry.

Low cost filters are frequently used. Slow gravity filters have the advantage of removing bacteria as well as suspended particles. Disinfection should be considered, however, if there is danger of bacteria, and chlorine pills, or sodium hypochlorite may be less dangerous than chlorine. Fluoride is difficult to control, and its use is not recommended.

REFERENCES

Abbott, J., 1988. Rural Water Supply. Continuing Engineering Education. A Course on Water Resources in Developing Areas. University of the Witwatersrand.

Economic Commission for Latin America, 1973. Popular participation in development in Community Development Journal (Oxford), Vol. 8, No. 3.

M.B. Consulting, 1987. Village water supply management handbook (A report published for USAID, Maseru).

Oakley. P. and Marsden, D., 1984. Approaches to participation in rural development (I.L.O.).

Stephenson, D., 1984. Pipeflow Analysis. Elsevier, 204 p.

Uphoff, N.R. and Cohen, J., 1979. Feasibility and application of rural development participation: A state of the art paper (Cornell University).

WHO, 1982. Activities of the World Health Organization in promoting community involvement for health development (Geneva).

CHAPTER 12

HYDRO ELECTRIC POWER DEVELOPMENT

INTRODUCTION

There are a number of factors supporting the development of hydro electric power in developing areas which come about from policies established for or independently of the necessity for development. There are however, a number of factors which disfavour hydro power development in these areas.

Although there are many developing areas which are supposedly arid, there are other areas which are most appropriate for developing hydro electric power. The rivers flowing to the east and south-east in South Africa have high discharges compared with the rest of South Africa, and these areas have not many other uses for the water. It would therefore, appear that hydro electric development could result in money to circulate in those areas, stimulating the economy.

ECONOMICS OF DEVELOPMENT OF HYDRO POWER

One of the factors which strangely encourages developing countries to develop hydro power is first world countries tariff structures imposed on these countries. On international grids, charges to neighbouring countries are equivalent to those of bulk consumers, but the latter includes certain amounts for distribution, transmission and administration as well as financing. On the other hand when a single country does an economic analysis it will compare its own costs with the prices which would otherwise be charged to that country,. and this tends to favour local development rather than importing electricity.

It is also possible to use tariffs in such a way as to minimize the payments by keeping purchased power on base load so that the energy is at a high load factor. Thus hydro power is generally more suitable on peak load where the power tariff includes an energy component and a monthly power peak cost.

Hydro power needs fairly large scale development, generally, to be competitive, however. That is, it is often difficult to justify micropower unless transmission costs are added in for that particular area only. This is generally not done, and organizations in particular charge a total tariff (including a large proportion of the transmission cost) wherever power is required.

Hydro electric power is in general capital intensive and requires large investments in the way of dams, tunnels and power stations before the project can be productive. It is, however, difficult for developing countries to get this capital as their revenue is insufficient to generate it. On the other hand, developing countries also are regarded as a poor risk investment in many cases so that they have to pay premium interest rates. All this tends to discourage hydro electric development. Economic justification is also highly susceptible to initial estimates, and often final costs exceed the initial estimate on which a scheme was justified. After that, difficult ground conditions encountered may increase engineering costs, floods may cause damage or bring silt, damages may render the scheme unpayable etc. Potential construction problems in developing countries are also highly unknown. Spares are difficult to obtain, and labour problems are more complicated. All these factors disfavour hydro power development when considered on an economic basis only.

A secondary argument in favour of hydro electric power and, in particular, dams and water development in these countries is that project construction will bring jobs to many people and, therefore, money to the country. Unless a deliberate attempt is made to do this however, generally the most competitive bids when the job is put out to tender are from mechanised contractors. Even if labour intensive methods are specified, the amount of money put into circulation is often short-lived, and after a few years the labourers are without employment again. Thus, it is probably not as effective as if a national training programme and development facilities for local contractors were established.

RISK IN POWER DEVELOPMENT

Apart from the fact that developing countries often have unstable economies that discourage investment (Cabora Bassa is an example of a big development which to some extent is wasted owing to the political scenario in Mocambique), there are a number of other risks to be considered in development of hydro power; the hydrology of many rivers in remote areas is highly variable and there is always a chance that an extreme drought will occur, leaving the country at the mercy of neighbouring countries for power requirements or else bringing the country to a standstill. On the other hand, this may be a reason for linking into a larger grid to prevent this happening.

USE OF LOCAL FACILITIES

The larger and more sophisticated a power station, the more likely it is to be designed by a country with a sophisticated infrastructure and technology, with the design fees retained outside the developing country's borders. The same often applies to contractors who (if highly mechanised, as required for many large projects) are often based at the larger centres.

The problem then arises as to how to involve the local people in order that expenditure for the project are circulated within the country. It is perhaps just as inefficient when expenditures are paid to designers and constructors outside the country as where electricity is purchased from an outside electricity corporation.

Local expertise should be employed at all levels of development of such projects, in particular at the planning stage. Local people should have a responsible role in planning, economic investment and technical factors in order to become familiar with the responsibilities and operation of the system in subsequent years. It is considered that construction time could be sacrificed, if necessary, in order to preferentialy use local people even if they have to be trained prior to being of use. It will also be necessary to train operators, managers and maintenance crew, and the availability of such staff needs to be considered at planning stage.

ECONOMIC ASPECTS

Interest rates applicable to the development of a dam or other major capital investment have an important bearing on the economic viability of the project. Current interest rates are relatively high, i.e. over 12%, compared with rates a few years ago, but on the other hand inflation rates are often higher. In fact, with present inflation rates the real interest rate could be negative although the development banks and other development sources have indicated rates between 6% and 3% should be considered. These are real interest rates i.e. the time preference rate for money which otherwise is static in value. It is the same as borrowing money from a country with a very low inflation rate at this interest rate.

It may be that developing countries should choose a lower real discount rate which would justify larger capital intensive projects in preference to operating intensive projects. This however could only be the case if the capital is spent within the country. Shadow values could be used however to distinguish between preferences. The rate of interest and the rate of inflation in developing countries should probably be assumed higher than for developed countries in this type of analysis, because development can often proceed at the expense of inflation, i.e. savings are used. Whether real interest rates, or financing rate plus inflation are used may not affect the results of an economic analysis, but it will affect a cash flow study. It is a shortage of cash which can affect the viability of large projects to a small power company.

CASE-STUDY-ECONOMICS (Taylor and Stephenson, 1986).

Based on cost of power purchased at 3 cents per kWh of energy plus approximately 2 cents per kW of peak hour power (converted from a monthly figure), a hydro scheme can be shown to be justified. The justifiable expenditure on hydro power assuming initially 100% load factor would be based on the amount saved in purchasing power.

Based on a thermal station cost of $1000/kW and fuel and operating costs of 1c/KWh, then at a load factor of 1.0 the present value would be $1000 + (1/100) × 8760 × 20 = $2750/kW, where 20 is the assumed present worth factor. The figure of $2750 represents the unit expenditure justified on a hydro station. On the other hand if the hydro station were operated at a lower load factor, for example, 10%, the justifiable cost would be R1175/kW installed. The cost of the dam to provide firm energy is often the overriding cost of a hydro scheme and as a result hydro power stations frequently operate only to meet peaks in power demands. (It should be noted transmission and overhead costs are omitted here).

The justifiable expenditure for storage is a function of the yield of the dam in m^3/s and head available. Thus if the yield in m^3/annum was equal to the capacity of the dam in m^3, and the head available was 100m, the energy available per m^3 of capacity would be $1m^3$ × 100m × 9.8/3600 = 0.27kWh/annum. The present worth of this energy at 1c/kWh energy cost would be 0.27c/annum = 5.5c present worth. Thus 5.5c could be spent per m^3 of reservoir capacity. For a 100 Mm^3 dam the justifiable cost would be only $5.5 million. Few dams can be built at this rate; i.e. storage is expensive in comparison with the benefits of energy. However, owing to increased depth, the benefit increases with scale.

Since hydro power is a non-consumable resource, i.e. water is available for other uses after generation, it seems there is no reason for not developing hydro power as fast as demand warrants. However, present tariff structures and surplus capacity make it difficult to export peaking power to developed countries.

In theory the high capital cost of hydro stations has the disadvantage of causing peaks and troughs in the economy. Labour is utilised largely only during construction. The high initial expenditure is also subject to risks. Even if the demand were to slump, the repayments would have to continue to be made, and if there was some mishap then repayments for power purchased from their system would become very high. Thus the planning of hydro power stations should be done very conservatively.

Another factor to consider is the reliability of the river flow. The estimates of the river flow should be conservative, and the variability of inflow from year to year should be closely studied. If for instance the flows were over-estimated, it may be impossible to meet the power demands if the hydro station were designed for peak loads only. During droughts the load factor of a hydro system should be decreased, and in this way the total power demand can still be met. During periods of high flow the hydro power plants could also be operated on base load.

CASE STUDY - COLLYWOBBLES SCHEME ON MBASHE RIVER

This scheme was commenced in 1982 and commissioned in 1984. It involved the construction of a diversion weir with a gross capacity of 9.5 million cubic metres and a 1.3 kilometre tunnel connecting down to the power station which is actually 30 kilometres downstream along the river. The station is equipped with three 14MW units, and the average operating head is 135 metres. It was planned to operate this station in a hybrid fashion, that is, at base load during summer and for peaks in winter with a limited storage installed initially. Fortunately the Ncora dam upstream discharges into a tributary of the Mbashe and provides some water during winter in addition to the low Mbashe flow of around 2 cumecs. The diversion from Ncora was increased by construction of an additional tunnel from the Ncora dam to the head-waters of the Mbashe tributary, but this bypasses Ncora turbines.

A number of problems occurred after commissioning of the Collywobbles Station due to inadequate estimate of floods (Stephenson and Collins, 1988) in the river, and the station had to be flood-proofed at great expense after major flood damage. A peak flow of 2 000 cumecs was recorded in February 1985. A subsequent flow of 3 000 cumecs occurred later in the year and 1 000 again in 1986. These flows are all above the original estimate of the 100 year flood. This emphasises the importance of correctly establishing the hydrology of the river. The result was that the entire hydrology and hydraulics of the system had to be revised, and emergency methods were taken to raise the walls of the power station by 3 metres. The sub-station was also raised and protected with gabions. A second problem subsequently developed at the Mbashe weir as a result of the deposit of silt behind the weir. Approximately 8.0 million cubic metres of silt have practically filled the reservoir and required drastic change in operational policy of the system in order to ensure the continued supply of electricity. The developing country was left to foot the bill for these problems, with a resulting stifling power charge.

METHOD FOR LOCATING OPTIMAL SITES FOR HYDROPOWER STATIONS

Mdoda (1986) described a method of identifying suitable hydro sites on rivers from the point of view of flow and head. To apply mathematical model analyses to hydropower development, assume that water flows into a turbine producing hydro-electric power P:

$$P = \gamma QH \tag{12.1}$$

where

γ = specific weight of water

Q = volume of water flowing into the turbine per unit time;

H = height between the headwater and the tailwater;

If the height increases by ΔH, P increases by ΔP at a distance L downstream on a site for hydropower station with the same discharge. From Eq (12.1) the increased hydropower is correspondingly:

$$P + \Delta P = \gamma Q(H + \Delta H) \tag{12.2}$$

Subtracting Eq (12.1) from Eq (12.2) gives:

$$\Delta P = \gamma Q \Delta H \tag{12.3}$$

Dividing Eq (12.3) by $(\gamma \Delta L)$ gives:

$$\Delta P/(\gamma \Delta L) = Q\Delta H/\Delta L \tag{12.4}$$

Taking the mean annual value q of river flows instead of the general value Q, Eq (12.4) gives:

$$\Delta P/(\gamma \Delta L) = q(\Delta H/\Delta L) \tag{12.5}$$

The ratio:

$$(\Delta H/\Delta L) = G \tag{12.6}$$

is called the "hydraulic gradient", From Eqs (12.5) and (12.6):

$$\Delta P/(\gamma \Delta L) = q \cdot G \tag{12.7}$$

The product of mean annual flow q and hydraulic gradient G is called the "hydro-electric potential" (P^{HDR}):

$$P^{HDR} = q \cdot G \qquad (12.8)$$

If A is the area of the river basin draining into the site for a hydropower station located a distance ΔL downstream of the preceding site, then from Fig. 12.1, a value of Mean Annual Runoff, $R^{MA} (= q/A)$, for the said site can be found. Then:

$$q = A \cdot R^{MA} \qquad (12.9)$$

and Eq. (12.8) gives:

$$P^{HDR} = A \cdot R^{MA} \cdot G \qquad (12.10)$$

For the k-th site for a hydropower station, the abovenamed quantities are assigned the index k (e.g. ΔH_k, is a water head measured from (k-1)-th site located on a distance ΔL_k upstream from the k-th site for hydropower station. From Eq (12.6), $G_k = \Delta H_k / L_k$, and Eq (12.10) gives:

$$P_k^{HDR} = A_k \cdot R_k^{MA} \cdot G_k \qquad (12.11)$$

To find values of A_k and G_k one has to use a topographical map of the region in question. For each location the data of latitude, longitude, elevation of river bed and possibilities for dam construction, are recorded. The total area A_k of drainage basin upstream of point is measured. The value R^{MA} is found by interpolation from Figure 12.1. The reduced level RL_k is found from topographical contours approximate to ΔH, as follows:

$$\Delta H = (RL_{k-1} - RL_k)/\Delta L_k, \quad k \geq 1 \qquad (12.12)$$

from which Eq. 12.6 (with index k) gives:

$$G_k = (RL_{k-1} - RL_k)/\Delta L_k, \quad (k \geq 1) \qquad (12.13)$$

For values of k > 0, the following values are tabulated: k, latitude, longitude, A_k, R_k^{MA}, RL_k, L_k. The corresponding values of G_k and P_k^{HDR} are computed (Eqs. (12.13 and 12.11)). The values of P^{HDR} are plotted on the watercourse of the corresponding river on the regional map (as shown on Fig. 12.2) from which the sites with optimal values of P^{HDR} can be identified.

Such a method is applicable to river systems in any region.

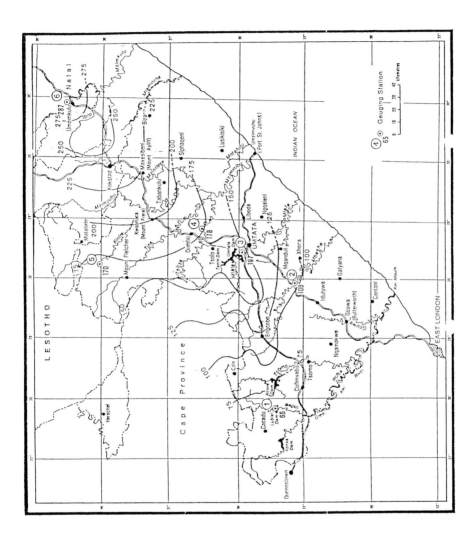

Fig. 12.1 Main Rivers in Transkei and mean annual runoffs (mm/annum)

Application of the Method

The planning method presented in this study was applied to major rivers in Transkei, namely: (a) Tsomo, (B) Mbashe, (c) Mtata, (d) Tsitsa, (e) Tina, (f) Kinira, and (g) Mzimvubu (See Fig. 12.1).

Values of RL_k (metres) were recorded at constant intervals $L_k = 1\ 000$ metres (1 km), and any sudden level changes could be identified in computations of G_k. However, to avoid excessive clustering of data for a point k, the values of P_k^{HDR} were computed for intervals of 10 000 metres (10 km). Thus, for Eq. 12.13:

$$G_k = [RL_{k-1} - RL_k] \text{ (metres)}/[10\ 000] \text{ (metres)}$$

$$\text{or } G_k = [RL_{k-1} - RL_k] \cdot 10^{-4} \text{ (m/m)}, \quad (k \geq 1) \tag{12.14}$$

The values of A_k in square kilometres (km^2) are measured from the map, and R_k^{MA} in millimetres per annum are interpolated from Fig. 12.1. Thus using Eq. 12.14, Eq. 12.11 becomes:

$$P_k^{HDR} = A_k \text{ (km}^2\text{)} \cdot R_k^{MA} \text{ (mm/annum)} \cdot [RL_{k-1} - RL_k] \cdot 10 \text{ (m/m)}$$

$$\text{or } P_k^{HDR} = A_k \cdot R_k^{MA} \cdot H_k/10 \text{ (m}^2\text{/annum)}, \quad (k \geq 1) \tag{12.15}$$

The units for contours of P^{HDR} are taken to be million cubic metres per annum (Mm^3/annum), since these are suitable for estimating the annual energy at the site for hydropower station.

Then Eq. 12.15 becomes:

$$P_k^{HDR} = A_k \cdot R_k^{MA} \cdot \Delta H_k/10^{+7} \text{ (Mm}^3\text{/annum)}, \quad (k \geq 1) \tag{12.16}$$

The values of P_k^{HDR} were computed and plotted on the map for the seven main rivers (see Fig. 12.2).

The results of work such as that of Mdoda can be used as a guide as to where to look for major hydro sites. For instance the existing sites on the Mbashe and Mtata rivers show high hydro potential. Even higher potential exist on the Mzimvubu and its tributary the Tsitsa. Feasibility studies have commenced for projects on these rivers. These studies will assess the hydrology and site dams in more detail and cost them. The power will be matched to loads in the most effective manner. Optimization

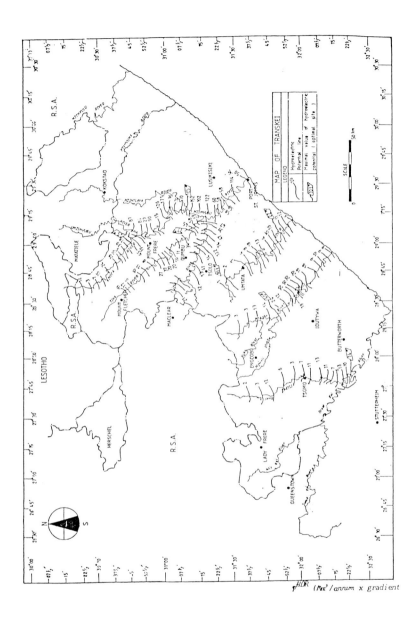

Fig. 12.2 Hydroelectric potential of river systems in Transkei

studies will be needed for this, and multi-purpose development of the rivers may become a possibility.

COORDINATION OF PLANNING AND DEVELOPMENT

It is important that limited water resources be used to their best advantage on a regional basis. Planning of developments such as hydro-electric schemes should, therefore, be done to realise the maximum potential of this limited resource. Planning should begin with a thorough basic study of hydrology and water resources. Many schemes have been and are being proposed with inadequate data. An integrated data collection system to develop water resources must be developed. Continuous records are required of the following:

1) Precipitation in the various catchments with both totals and rates.

2) River flow data including variations from season to season, year to year and peak flood flows.

3) Sediment loads in rivers. This again will have to be on a continuous basis as concentration of silt depends very largely on the river flow rate.

In parallel with the hydrological investigations, an in-depth analysis of local and regional needs is required. This includes:

4) Local Needs, viz:

   4.1 Local power and energy requirements.
   4.2 Local water requirements to meet existing and proposed agricultural, industrial, commercial and domestic developments.
   4.3 Environmental requirements and

5) Regional Needs Analysis. Sensitive consideration of neighbouring countries' electricity and water supply needs and economics is needed.

Once resources are identified and needs covered, both local and regional, then potential developments should be classified within the regional or local context. Detailed objectives for the evaluation of proposed developments need to be stated and guidelines established. Expansion of hydro-electric facilities should then take place within this framework. In general ad hoc proposals could spoil the ultimate plan and even not meet immediate demands in the most efficient manner.

SMALL-SCALE HYDRO PLANTS

There has been considerable interest in small hydro-electric plants. The cost of such plants is often justified by the savings in transmission costs. Thus mini plants ($\sim$2kW) may serve rural communities or industries and micro plants ($\sim$500W) could serve individual houses away from other sources of electricity. The civil and mechanical costs per unit of such plants are, however, generally higher than for large-scale plants, and this prohibits their use for rural and poorer communities in particular, the very people they are often intended for.

Use of pumps in reverse has been advocated by Dutkiewicz (1986) but again such applications are confined to use on suitable streams where there is adequate water, as mechanical efficiency is then not important. Efficiencies of 50-80% can be expected, compared with over 90% i.e. for large plants.

Small scale hydro power development is not new. It has been used for centuries for driving mills and more recently for electricity generation for specific purposes e.g. a factory. With the advent of subsidized rural transmission systems from larger plants to meet more general electrical demands such small scale plants become redundant.

The energy crisis of the 1970's revived the interest in small scale hydro power (defined as less than 25kW). The linking of small plants to national grids has become relatively easy; however, this requires constant speed of rotation of the generator at a synchronous speed. The buy back price paid by electricity authorities is generally less than their selling price as the energy is not regarded as reliable.

Generally small plants do not have storage i.e. they are run of river, because the flow requirement is well below the flow of the river.

If the small plant is only for a local use, low voltage distribution is often economical, compared with the necessity for very high voltage over long distances to minimize volt drops.

Not only rivers provide hydro-electricity. Frequently canals with falls, or outlets to irrigation areas, can generate small heads. Such may be suitable for reversed flow pumps acting as turbines. Alternatively, pumps may be installed to provide water supply to higher lying villages, or to provide water for a mini pumped storage arrangement with higher head

246

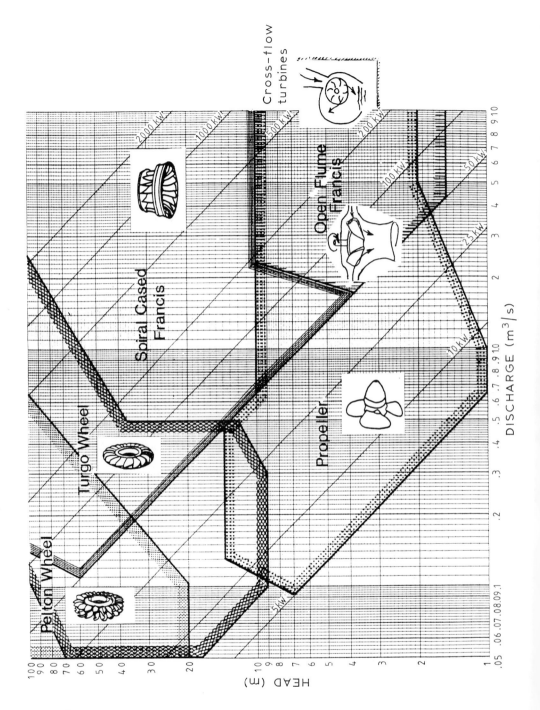

Fig. 12.3   Range of application of various turbine types.
(University of Salford, 1983)

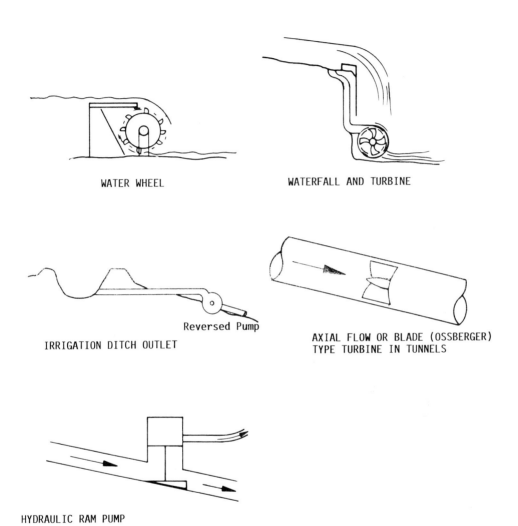

WATER WHEEL

WATERFALL AND TURBINE

Reversed Pump

IRRIGATION DITCH OUTLET

AXIAL FLOW OR BLADE (OSSBERGER)
TYPE TURBINE IN TUNNELS

HYDRAULIC RAM PUMP

Fig. 12.4 Micro hydro options

and ability to generate as called on.

Intake screens are an important aspect of small as well as large hydro plants. Debris can damage turbines and blockage of a turbine could cut off power supply, with the loss of reliability in the system.

Selection of pipe sizes is another area requiring careful engineering as pressures can be high and friction losses may reduce initially envisaged outputs.

Research into using low cost pumps in reverse is in progress to reduce costs of small turbines. In addition, novel distribution systems merit study, e.g. low voltage uninsulated wires to local houses or villages.

Machine Selection

The head and flow rate will determine the most efficient type of hydraulic turbine. Various types of machine are listed below, but more detailed information must be sought from books or suppliers see (Fig 12.3)

Impulse turbines: water head is connected to velocity through a nozzle, and drives a *Pelton wheel* with buckets (heads up to 200m) or a *Turgo wheel* with blades, (heads up to 20m) or a *cross flow* turbine with curved blades. The latter are particularly viable and run with flows down to 16% of design flow rate. Heads from 2 to 100m can be used.

Reaction turbines: have a pressurised casing, and include the *Francis turbine* with radial flow (operating at heads from 30 to 500m) and the propeller type turbine with axial flow (operating at heads from 3 to 60m). Centrifugal pumps in reverse operate similarly to Francis turbines.

NON-ELECTRIC HYDRAULIC POWER

Whereas we tend to regard electric power as the cheapest and most efficient in our homes, this is not always the case. In rural surroundings priorities differ. Priorities for lighting and heating may be secondary to pumping water or grinding. In such cases mechanical converters may be all that is required. Hydraulic ram pumps (Fig 12.4) are low cost alternatives for water pumping. Water wheels may be used to grind grain. Gravity flow in canals may also be more reliable than pumping even though it may limit the use of water. Such alternatives need social consideration before the engineer rushes in to provide his idea of power.

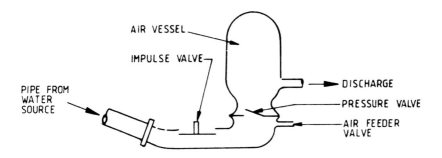

AIR VESSEL

IMPULSE VALVE

PIPE FROM
WATER
SOURCE

DISCHARGE

PRESSURE VALVE

AIR FEEDER
VALVE

Blakes hydraulic run 32m bore – 1 000ℓ/day

Supply Head m	Delivery Head m					
	5	10	20	50	100	150
1	1.4	.65	.29	.08		
2		1.5	.79	.25		
5			2.3	.94	.36	.12
10				1.9	.93	.4
20					1.8	.7

Fig. 12.5 Diagram showing essential parts of Hydraulic Ram

REFERENCES

Dutkiewicz, R.K., 1986. The potential for small-scale hydro plant. Energy
    Research Institute, University of Cape Town.
Mdoda, G.M., 1986. Method for locating sites for hydro-electric power
    stations. Proc. Conf. on Renewable Energy Potential in South Africa.
    University of Cape Town.
Stephenson, D. and Collins, S., 1988. Problems due to innacurate flood
    estimates at Collywobbles. ICOLD, San Francisco. Q63, 691-701.
Taylor, R.W. and Stephenson, D., 1986. The appropriate development of the
    hydro-electric resources of Transkei. Proc. Conf. on Renewable Energy
    Potential in South Africa, University of Cape Town.
University of Salford, 1983. Department of Civil Eng. Report on The
    Development of Small Scale Hydro-Electric Power Plants.

CHAPTER 13

HUMAN RESOURCES

STAFFING NEEDS

Available information indicates a critical shortage of specialists in water resources, at both professional and subprofessional levels, in many countries. Educational programmes are needed to gradually eliminate this shortage and to ensure that both current and foreseeable problems associated with more extensive development of water resources can be solved. Staffing problems widespread in water resource programmes in developing areas are rooted in a shortage of skilled staff, large numbers of untrained staff, managers without sufficient training and autonomy, political appointments, and weak accounting and financial systems. Poor maintenance and periodic breakdowns are more frequently due to untrained staff than to other reasons, and training of staff is frequently neglected. Hydropower plants, dams, water supply, and related facilities, for example, have to function twenty-four hours a day for the life of the facility. Few developing areas have the trained manpower required for such operation.

Education and training programmes in water resources have to be adapted to the economic and cultural conditions in a given country or region, and the degree of specialization in such programmes decreases from graduate to undergraduate and from professional to subprofessional levels. Development of adequate national education programs is often hindered by the lack of qualified teachers and facilities. Additionally, there is a need for awareness by planners and decision-makers of the role of water resource development in overall socio-economic development of a country, and appropriate educational programmes are needed to meet this need.

In lesser developed areas the growth of the human population and the desire for an improved standard of living have resulted in unprecedented need for development of water resources to provide water supply for irrigation, domestic, and industrial use, generate hydroelectric power, and so on. Growing water scarcity dictates that available resources be utilized carefully, be protected from pollution, and be conserved to the utmost extent especially in arid areas.

The basic target of education and training programmes in water resources in developing areas often is to ameliorate the adverse effects of climate, drought, and water scarcity so that self-sufficiency in food production can be achieved in both the short and long term and to improve economic conditions and quality of life. The specific objective is to assist those countries in developing a cadre of engineers, scientists, specialists, and technicians in engineering, agriculture, and other areas related to comprehensive development and management of water resources.

Various United Nations studies indicate that the supply of local professionals and technicians in lesser developed areas needs to be increased significantly in the near future to meet needs. The most important problems identified with education and training in those studies include the following:

1. Education and training programmes in water resources must be adapted to the cultural, social, economic, and environmental conditions in a given country.

2. Development of adequate local educational programs in water resources is often limited by the availability of qualified teaching staff and facilities.

3. One of the most crucial problems in water resources development is with competent operation and maintenance. Training of a corps of professionals and technicians for operation and maintenance should be of high priority. Even in countries subject to administrative changes the basic data preparation can proceed.

Because problems, needs, and resources differ significantly from country to country, there is need for country-specific programs for education and training based on:

1. Identification of education and training needs.

2. Evaluation of human, technological, physical, and financial resources at the regional, national, and international levels to meet the needs.

Usually developing areas have need for both complex water development programmes and for simpler single-purpose increments. Primarily due to economic efficiency, there has been a trend towards more complex water programmes that include a number of different facilities and have multiple objectives (not all of which are compatible). Sophisticated operating criteria may be required to realize optimum benefits from such programmes, and management for the total system may be shared by several ministries or agencies. The characteristics and training needs for operators of complex and of simple systems are very different. However, basically all systems must interact with the intended beneficiaries at the users' cultural level.

The administrators, managers, and technicians who operate complex water resource systems must be trained in such a way that they can reduce the complexity of the system to manageable levels. It is also important that complex systems are operated efficiently through exchange of information among the various agencies typically responsible for managing water. Training for these tasks requires a structured set of teaching methodologies that leads effectively from the training phase to operating projects.

CONSULTANTS

The use of qualified local staff for conducting water resources studies offers a number of advantages. They have superior knowledge of social customs, local environment, and institutional and legal constraints than do expatriates. For some activities, such as contacts with rural people, local staff who speak a common language or dialect is essential. Further, building up a national capability to conduct feasibility studies and implement programmes is an important national development objective in itself. Developing local capability to conceive, design, and carry out projects is an important part of a country's development process; otherwise, a country cannot fully control its economic and social development.

Local staff, however, are not usually able to perform highly specialized studies, and experienced qualified foreign consultants are often needed for such work. Consultants are commonly used in water resource project work for;

1. Preinvestment studies that normally precede a decision to go forward with a project, including feasibility studies.

2. Preparing design documents required for invitation of bids for construction.

3. Construction supervision and project management.

4. Technical assistance, including a wide range of advisory and support services, such as national and sector planning, organization and management studies, staffing and training studies, and assistance in implementing study recommendations.

When possible, these tasks should be conducted by local staff because of their understanding of local conditions and insight. Also, selection or development of technology appropriate to local conditions is facilitated when local staff is responsible for these tasks.

Few agencies in developing areas have staff versed in all the areas of specialization necessary for project work, but this capability can be developed through the educational system, on-the-job training, or study abroad. Initially, it may be necessary for a country to choose between building up local capability and expeditious design and construction of high priority programmes. The World Bank in its lending programme, has found that many countries, beginning with relatively simple projects, have progressively strengthened the capability to design and construct their own projects with little outside help.

When advisers are needed, there are definite advantages in using local consultants because of their knowledge of local conditions and customs. Costs are likely to be lower than using expatriates, and a greater proportion of the costs would be incurred in local currency. However, only a few developing countries have local consultants capable of providing the wide range of services needed for large, complex projects. There are also some occasions when the best choice is a foreign consultant because they are more independent of local political pressure and because they may introduce technologies with which the local staff is not familiar. The use of foreign consultants from the donor country is often specified by the donor, and this benefits both donor and receiver countries.

TRAINING AND APPROPRIATE TECHNOLOGY

As discussed earlier, it is basic that the technology associated with water resource development be appropriate for the users and for local conditions and culture as well as being technically sound. The level of technology associated with development programs need not be either the traditional technology of the local area or the most advanced technology available, but it must be at a level with which the intended beneficiaries can interact and which will lead to improved health and living standards as well as improved economic conditions. What is technically feasible is not necessarily justified in terms of either economic or social costs. The scarcity of trained supervisors is sometimes a constraint on the feasibility of labour-intensive methods, and a period of time is needed to build up a corps of skilled supervisors in a developing area.

In some cases is may be desirable to modify whatever technology is initially adopted over time, in stages, as users understand how each successive modernization will benefit them and are willing to make the needed changes. The principal purpose of education and training programmes is to transfer "real" technology. If the technology involved is too sophisticated or unacceptable on cultural grounds, it will not be accepted and used for any length of time.

In some societies how water is used involves complex cultural traditions, family practices, social interaction, and religious beliefs. Some programmes for water supply and sanitation have not been acceptable to users for these reasons, but also there is often a lack of understanding of the relationship between low standards of personal and household sanitation and disease. Traditional water sources are often perceived by users as being better than a safe new supply. Consumers must understand the importance of using safe water supplies and safe sanitation measures and the importance of keeping them safe. The only way this understanding can be achieved is by involving the users from the earliest stages of project planning. (Since women usually control the use of water within households, getting them to participate is especially important.) There must be individual commitment to the use of new technology and sustained support by local groups.

Many advanced technologies in industialized countries for domestic water supply and sanitation, for example, are designed primarily for user convenience – it is accepted that such supplies and facilities are safe from a health standpoint. There are other, less sophisticated means of supplying safe water (standpipes and courtyard connections, for example) that are more affordable and more appropriate for developing areas. The potential for later upgrading standards of service over time should not be overlooked.

Further, in some countries legal restrictions may be a problem in adopting affordable technology. At the time some countries achieved independence, they adopted codes of the colonial powers that are excessively restrictive. It may be necessary to modify legal constraints in order to utilize appropriate technology.

It is not always easy to identify the most appropriate technology. There is the danger that (1) local staff educated abroad may require a bias in favour of advanced technology or conclude that the latest technology is best; (2) expatriate consultants may advocate technologies with which they are most familiar; and (3) lack of communication with intended beneficiaries may result in errors in identifying real problems and needs and capabilities for adapting to new technologies. Foreign study programs sometimes offer little that is relevant to the needs of the trainees' country, and it may be difficult for them to adapt what has been learned abroad to local conditions. All these factors complicate selection of the appropriate level of technology for a given programme in a given setting.

Fields of Study and Training

The knowledge and skills typically most essential in developing areas in the near future are in the following broad areas:

1. Applied water resource planning, development, management, and administration.

2. Atmospheric science, hydrometeorology, and hydroclimatology.

3. Engineering and social and environmental sciences related to water resources, food production, and hydropower development.

4. Construction methods, supervision, scheduling, and cost control.

5. Operation and maintenance of water development facilities.

Study and Training Programmes

A number of different types of study and training programmes could be established to educate and develop a core group of faculty, engineers, planners, managers, scientists, construction personnel, and technicians to implement water management programmes, as follows:

1. Specially developed university-level programmes taught by expatriate specialists and concentrating on knowledge and skills needed locally and leading to advanced degrees by local institutions. Such programmes would be primarily oriented to train the large number of specialists and managers needed for a comprehensive water management programme.

2. Formal university-level ˙courses both locally and abroad leading to advanced degrees, primarily orientated to development of a key group of future leaders in water resource policy, planning, engineering, development, and management. Such courses would include related scientific fields such as climatology, economics, social sciences, environmental, public health, public involvement, and administration.

3. Workshops, short courses, and special lectures or training courses locally or abroad, as appropriate, leading to examinations and certificates of completion, covering principles and techniques that are suitable (or can be adapted to be suitable) for use in the local area, primarily for specialists and technicians.

4. On-the-job training locally and abroad, as appropriate, leading to examinations and certificates of completion, covering principles and techniques that are suitable (or can be adapted to be suitable) for use in the local area, primarily for specialists and technicians.

5. Study tours of institutions, projects, and programs that would contribute to better understanding of policies, laws and regulations, procedures, and technologies used elsewhere and thus strengthen capability in water resource project identification, conceptualization, implementation, and operation. Primarily for engineers, planners, managers, scientists, and construction and operating personnel.

Education and Training Course Subjects

Preliminary assessment of the knowledge and skills needed to develop human resources to implement comprehensive water resource development and management programmes using appropriate technology include the following:

<u>Planning and Design</u>

1. Fluid mechanics and engineering hydraulics.
2. Basic data acquisition and management – remote sensing, aerial photography, photogrammetry, land surveying and mapping; geologic surveying and mapping.
3. Remote sensing to assess land and water resources on a large scale and to monitor and assist in predicting long-term changes in

resources.

4. Forecasting – storm analysis; data collection; reservoir operation; forecasting temperature, precipitation, flash floods, streamflow, and snowmelt runoff.

5. Dendrochronology – techniques, reconstructing past hydrologic events, climatic variability, and drought cycles.

6. Drought – low-flow analysis; drought frequency, extent, impacts, and amelioration.

7. Advanced surface water hydrology – selecting gauging stations; field measurements; field instrumentation; data transmission, evaluation, storage, and retrieval; probability, return periods; probable maximum precipitation; infiltration; unit hydrographs and streamflow; flood routing; reservoir yield; evaporation.

8. Ground water hydrology – groundwater geology, acquifer characteristics; resource assessment; unsaturated and saturated flow; flow nets; acquifer tests; walls; land subsidence; contamination; modelling; ground water management; conjunctive use; artificial recharge.

9. Wells and boreholes – single wells, well fields, safe yields, overdraft, flow modelling, well drilling and completion; salinity intrusion in coastal aquifers; safe drinking water and public health.

10. Open-channel flow – channel controls; water surface profiles; critical flow; uniform flow; gradually varied flow; flood routing; time of travel; unsteady flow; water quality.

11. Sediment – stream transport, erosion, and deposition; watershed erosion; deposition in reservoirs.

12. Water resource planning and development:
   a. Policy.
   b. Identify objectives and goals.
   c. Legal and institutional requirements.
   d. Framework planning; river basin planning; project planning.
   e. Identify "most probable future" and do a risk analysis for alternatives.
   f. Evaluate existing dams and other water management measures (dam safety, spillway adequacy, sediment deposition and remaining useful storage, power potential, and so on.)
   g. Assess potential to improve quality of life for rural people by increased food production, improved nutrition, safe drinking water, reafforestation, hydroelectric power.
   h. Identify and evaluate feasibility of various water management

measures (additional large and small storage reservoirs; conjunctive use of surface and ground water; interbasin water transfers; increased fresh-water yield using protective measures in coastal acquifers; augmentation of supplies).

i. Develop alternative plans and assess economic, social, and environmental impacts, benefits, and costs.

j. For large-scale programs and projects

- Identify elements of comprehensive long-range water management programs; develop river basin plans; establish priorities; project planning.

- Select the "best" plan; evaluate plan elements in meeting national objectives, economic feasibility, impacts on public health and social well-being, and environmental effects.

- Cost allocations, cost sharing, and repayment.

13. Reservoirs (single purpose and multiple purpose) - storage capacity needed and available; operating rule curves; flood routing through reservoirs; sediment deposition; water quality; hydrograph modification; spillway adequacy.

14. Dams - types; site selection; hydraulic components (spillways, outlet works, gates and valves, terminal structures); design floods (diversion, cofferdams, project design, spillway design); environmental and social impacts of dams and reservoirs.

15. Dam - design, concrete dams, earthfill and rockfill dams, roller compacted concrete dams, dam safety (failure, spillway adequacy, dam-break analysis).

16. Water supply augmentation and conservation in arid and semi-arid areas.

a. Water harvesting.

b. Artificial recharge.

c. Reuse of waste water and brackish water.

d. Pipelines and lined canals for irrigation water supply.

e. Irrigation return flow systems.

f. Increased efficiency in use of irrigation water and modification of agricultural practices.

g. Control of phreatophytes and other weeds to reduce losses.

h. Evaporation suppression techniques.

i. Precipitation augmentation.

j. Limitations on pumping in areas of short supply.

k. Conjunctive use of ground and surface water (using ground water to meet peak seasonal demands and in times of surface water deficiencies).

17. River engineering – river morphology, basic data, stabilization and rectification of rivers, drop structures, dredging, diversion and cofferdams, levees, physical hydraulic models, impacts, benefits, costs.

18. Hydroelectric power (large and small plants).

  a. Site evaluation

  b. Types of development.

  c. Components of projects

  d. Types of turbines.

  e. Need for power and load forecasting.

  f. Hydrologic and hydraulic studies.

  g. Energy generating potential.

  h. Turbine selection.

  i. Powerplant sizing.

  j. Environmental and social impacts.

  k. Benefits, costs, and economic and financial evaluations.

19. Improved agricultural practices related to water – improved water measurement and control; incentives to improve irrigation water use; improved land levelling techniques; increased on-farm efficiency in water use; reduced seepage, evaporation, and transpiration losses; drainage; cropping changes.

Construction

1. Job scheduling (critical path).

  a. Time required for site access, for delivery of materials, for completion of project increments, and so on.

  b. Impact of climate and season of year on construction schedule.

  c. Equipment needed (type, quantity, duration).

  d. Manpower needed (number, when).

  e. Estimated expenditures over time.

  f. Time to complete project.

2. Project control during construction – progress reports, equipment records, cost records.

3. Tools, machines, and equipment (appropriate technology for locality) – tractors, scrapers, excavating equipment, trucks, conveyers, compressed air, drilling and blasting equipment, tunneling, grouting, pile driving equipment, pumps, stone crushers, forming for concrete and so on.

4. Concrete – mixture design, handling and batching aggregate, mixing plants, handling and transporting concrete.

5. Supervision of construction.
6. Inspection and inspectors reports.
7. Environmental controls – water pollution control; air pollution; dust; damage to vegetation, habitat, fish and wildlife, cultural resources.

Operation and Maintenance

1. Appropriate technology.
2. Simplified operating criteria understandable to local administrators, managers, and technicians.
3. Trained local operators and support personnel (mechanics, electricians, and so on).
4. Standardization of equipment and a reasonable inventory of spare parts.

Water Resource Management

1. Establish priorities and make timely decisions to construct and/or implement sequent elements of a comprehensive water management program or specific small-scale programmes.
2. Evaluate system performance and impacts over time.
3. Effective maintenance.
4. Modify master plan (construction and implementation of future programme elements) over time to meet changing needs.
5. Importance of realistic and fiscally responsible program of exenditures.
6. Need to modify allocation of water for various purposes over time if national priorities and needs change.
7. Monitor effects of water development and management on public health and social well-being over time.

LABOUR-INTENSIVE CONSTRUCTION

Investment in developing countries can take many forms and comes with many catches. Some foreign aid may be on the basis that the financing country must provide the consultants and contractors; the money doesn't really leave the donor country, and the developing country still has to pay the bill.

Less stringent is the provision of aid with no strings attached but overviewed and designed by professionals from developed countries who think in first world terms. Implicit in many such designs is that the solution should be 'least cost'. This may end up by causing draining of foreign reserves because the design needs machines and technology. While there is aid coming in, the value of the local currency may be exaggerated, and hence local labour use appears unattractive. The fact is that, whenever there is unemployment, any use of local rather than imported methods is valuable. Labour intensive methods are well suited to 'capture' the aid locally instead of re-exporting it.

Admittedly labour intensive construction is not the final or complete answer. The labour force can become redundant and finally dissatisfied when construction is complete. Some form of training and establishment of local enterprise should therefore accompany the use of labour intensive methods so that there may be impetus left in the developing country to generate further development and wealth.

Labour-Intensive Public Works Construction and Private Contractors

Labour intensive networks are being used for various types of construction work in sub-Saharan Africa. National programmes of labour-intensive rural construction and maintenance now exist in Kenya, Malawi, Botswana and Lesotho. The scale and scope of these programmes have been described elsewhere (de Veen 1980, McCutcheon 1983 and 1988).

All of these programmes have been implemented by the public sector, the agency normally being a works-oriented government ministry with no profit incentive. Labourers have been hired on a casual daily paid basis. On the maintenance programmes the labourers responsible for a certain project are termed both 'lengthmen' and 'contractors'. The latter term is somewhat misleading as the labourers are really self employed individuals, not formal construction companies. Except for one programme in Ghana, privately owned contracting companies have not yet played any part in the large scale development of labour-intensive methods of construction.

Against this background recent developments in Southern Africa mark a significant point of departure. In some African countries private contractors using labour-intensive methods are constructing certain public sector works. Only a few years ago this would not have been thought sensible in a country as technologically advanced as South Africa. The development is the outcome of several factors; an abundance of labour in the rural areas and an acceptance by senior policy makers of the need to create employment; the 'privatisation' drive at policy making level; the depth of the recession which has forced contactors to accept terms which would previously have been unpalatable, and a national awareness of a need for labour employment for stability. In fact it may be primarily the latter, which is forcing South Africa ahead in this field.

Below we deal with three aspects of the use of labour intensive methods by private contractors engaged on public sector work. Firstly, in order to substantiate the policy-makers wisdom in seeking to encourage the use of labour-intensive methods, mention is made of the soundness of the intellectual base. Secondly, the way in which public authorities are involving private contractors is described; thirdly, some aspects which require further attention in order to ensure the success of future ventures and finally major pitfalls to be avoided are described.

1. Definition of labour-intensive construction

   The productive employment of labour in the creation of a product that is technically and economically efficient: the product will be of as high a technical standard as the specification and the funding will allow.

2. Scope of work

   Agriculture, construction.

   Within civil engineering: road construction and maintenance; water supply (dams, weirs, canals) and drainage.

3. Unemployment

   Most projections of future unemployment are very high. Often worse for the rural than urban areas, but this receives less attention.

4. Employment creation

   Many remedies have been proposed. Sometimes 'labour-intensive' employment creation has been put forward as a panacea. Such a contention not only claims too much, but also leads to false expectations and often to programmes that are incorrectly implemented and thus tarnish the image. It should rather be said that proper labour-intensive programmes could make a significant contribution. While we are dealing with labour-intensive construction, it must be remembered that in the rural areas the greatest potential would be from an increase in the use of labour in agriculture.

5. Intellectual base

   Labour-intensive work is held to be backward and retrogressive. However there is a sound intellectual base for the increased use of these methods.

   In the late 60s and early 70s interest in the increased use of labour in civil construction grew out of:

(i)   the failure, in the Third World industrialisation policies of the 1950 – 1960s to achieve the 'take off' predicted by economists.

(ii)   recognition of the vast numbers of unemployed people in developing countries.

(iii)   the need to use local resources and save on foreign exchange.

Broadly speaking the technical and economic efficiency of the substitution of labour for equipment is partly dependent upon whether the use of equipment is essential: i.e.

(i)   modern methods require tolerances which could not be achieved by earlier methods.

(ii)   smooth succession of events from raw material to completed product.

(iii) final product not affected by final location.

(iv)   scale of production : cost of storage increases by square while volume increases by cube.

Looked at from this perspective civil construction was promising: the products were time honoured, the machines were magnified versions of simple tools, and the process of production was littered with possibilities for hiatus.

In much civil construction, earthworks – excavation, loading, haulage, unloading, spreading – typically account for up to 50 per cent of expenditure. Other promising operations are production of aggregates and pavement construction.

In 1971 the World Bank initiated a programme of work to explore this question in detail in relation to one section of civil construction; it investigated the technical feasibility of the substitution of labour for equipment in road construction. Significantly its first interim conclusions were:

It is technically feasible to substitute labour for equipment for all but about 10 to 20 per cent of total road construction costs for the higher quality construction standards considered, while relaxation of standards to an intermediate quality permits labour substitution for only an additional 5 to 16 per cent of costs (i.e. for all but about 2 to 15 per cent ot total cost).

This was later generalized to say that labour-intensive methods were technically feasible for a wide range of construction activities and could generally produce the same quality of product as equipment-intensive methods.

Having proved the technical feasibility, the World Bank proceeded to deal with questions of economic efficiency. The World Bank's 1983 Sector Support Strategy Paper for Transportation included the following in its summary:

Greater efforts must be made to spread the use of labour intensive techniques of construction and maintenance, whether for transport or other purposes. Where-ever the basic wage actually paid...is less than... about US$4.00 per day in 1982 prices, and labour is available in adequate quantities, the alternative of using labour-intensive techniques should be seriously considered.

JUSTIFICATION OF THE USE OF LABOUR-INTENSIVE METHODS

The technical feasibility and economic efficiency of the substitution of labour for equipment in civil construction, particularly earthworks, was theoretically established during the nineteen seventies (IBRD, 1971, 1974, 1983, 1986; Deepak Lal 1978). Since then it has been amply demonstrated by the practical implementation of various programmes.

For small scale low standard road construction projects in remote areas, the use of labour-intensive methods can compete on a financial basis with alternative equipment-intensive operations. The theoretical contentions that labour-intensive work could also be used on high standard construction has not only been demonstrated in Asia but also in Africa i.e. in Zimbabwe (International Labour Organization, 1988). However, for high standard construction near industrialised areas (i.e. easy access to equipment, spares, fuel, mechanics and workshops) it is necessary to use economic considerations (shadow prices/wage rates, foreign exchange) to prove that labour-intensive methods are economically efficient when viewed from the perspective of the nation as a whole.

One of the crucial outcomes of such an analysis must be the specification of the wage that should be paid to unskilled labour. This wage will reflect a compromise between the shadow wage rate, the sum actually needed to provide a reasonable livelihood (itself dependent upon the price of essential goods, the average size of a household and the average number of breadwinners per household), the average wage being paid to unskilled labourers in the private sector, the average wage being paid to unskilled labour in the public sector and the relative power of other concerned bodies such as the trade unions, the employers' federation, and the government ministry responsible for labour and its regulation.

If the resultant wage is low and labour efficient, it may be found that labour intensive methods can compete financially with alternative methods, in which case the government should not need to intervene. However, if the resultant wage is too high for labour-intensive methods to compete financially, the government must either reduce the wage to suit itself or accept the higher short-term burden on the exchequer, There are different options as to how to do this, the simplest being to put the work out to contract. In Southern Africa a compromise has been reached; a national minimum wage has been specified for this type of work and funds have been made available by government to stimulate such work; at the same time contracts are put out to tender.

PUBLIC AUTHORITIES VERSUS PRIVATE CONTRACTORS
(McCutcheon and Stephenson, 1988)

To date public authorities seem to be using one simple method of securing the use of labour-intensive methods by private contractors: contracts specify that labour-intensive methods should be used. Contractors have to tender accordingly. Loose sections in a document such as above could be ignored or misused in many ways by a profit orientated contractor who will use labour to the extent that suits himself. Contractors have increased prices by up to 25% to account for the inconvenience of using unskilled labour.

Several refinements to this simple contractual approach have been noted. One is to break the work into different categories and specify the type of equipment that may be used on such operations. Another is to specify that the contractor is responsible for any training that is necessary in order for labour-intensive methods to be used. The contractor will have to be paid for the training.

A third is to specify the use of sub contractors with a certain level of skill. The latter offers many possibilities. However, administrative costs, purchase or lease of plant for small contractors (e.g. even a light truck) all cost something. Training in the technical and managerial skills, and promotion or growth also require money. Low interest loans could be used for this.

Wage Rate

Above it was stressed that the wage rate for unskilled labour was a critical factor. It is doubtful whether sufficient work has yet been carried out to ensure that the current wage rate represents the best compromise between the competing factors. If the rate is too low, on the one hand it will be exploitative, demoralising and give labour-intensive work a bad name from which it may never recover, while, on the other, the work force will be insufficiently motivated to achieve high productivities (and quite rightly so). If the rate is too high, ultimately someone will raise the question "what is the worth of the work for which we are paying good money?"

Specifications and labour-intensive methods

While it is a good start to specify the type of equipment that may be used for certain categories of work, it is questioned whether this will achieve the right objective. Specification of ostensibly simple road construction equipment in the hope of creating employment opportunities has been known to result in an essentially capital-intensive operation with gangs of labourers watching the equipment. More importantly, the extent to which labour-intensive methods can be used for almost all operations in civil construction is probably not properly recognised. If the engineer/planner in the public authority (or the consultant to the public authority) is not aware of the labour-intensive alternatives (or does not trust them because of lack of experience with them), the specifications themselves will not be suitably prepared.

It is a case of educating the clients and their advisors. As a first step we suggest that every public authority should obtain the ILO's *Guide to Tools and Equipment for labour-intensive work,* and as many as possible of the World Bank and ILO publications on the subject.

Training

While it is a good ploy for public authorities to insist that contractors be responsible for training, perhaps this should be augmented by some form of field training school which will teach fundamental principles and techniques and from which contractors can draw personnel. The contractors will continue to train, but from a higher base than at present; this will enable contractors to concentrate on what they are best at doing; practical construction.

If training in labour-intensive methods is left entirely to contractors, we should not be surprised if such training is quietly lost to the nation as a whole. The same contractor cannot be guaranteed to win the next labour-intensive contract, and the contractor's next contract may be entirely capital-intensive.

Labour only sub-contractors

For an entrepreneur a major advantage of labour intensive work is that very little capital is required to embark upon a venture. A major disadvantage for the client authority is the danger that the ease of entry with this type of work, could attract not only the inexperienced but the unscrupulous. As to the latter, business credentials should be scrutinised carefully. As for experience, there is a chicken and egg situation; without prior experience one may not be able to get work, without work one cannot get experience. Here, a combination of labour-only sub contracts and a careful record of serious failures should enable the prospective entrepreneur to enter the industry and yet protect the client authority. Under these circumstances, it would be mutually beneficial for the training school which produces skilled site operators and supervisors, to also conduct elementary business courses for fledgling entrepreneurs.

Under conditions where the work is mainly carried out by labour-only sub-contractors, the major contractor must adopt the role of the overall construction manager.

ECONOMICS

The fact that labour-intensive methods do not appear economically attractive now could be because mechanical methods are artificially cheap; i.e., the currency could still be overvalued, making imports of equipment appear economic. The 'high' currency could be supported by commodity prices but it could fall until labour is competitive with mechanized methods. Then imports would fall, stabilising the currency.

Until such time (if the theory is correct) the price of labour may not reflect its true worth from the national point of view. There is a definite economic value in providing employment, in using local under-utilized skills and in training. Such value must be decided at a national level. Shadow pricing is one way of putting value on commodities which are required for national development. Labour should be priced at its shadow value, which could be zero if the labour was otherwise unemployed, even though labour is actually paid a wage. A method of establishing shadow values is possible using a planning model. Such a computer model, which uses the principle of decomposition of linear programmes, is described in Chapter 5 (Stephenson and Paling, 1987).

The price submitted by a contractor at the time of tender, should on this basis be adjusted before adjudicating (tenderers should however be made aware of the basis). Labour rates should be adjusted by the shadow value. It may therefore be necessary to price tenders on an activity basis instead of item basis to enable this to be achieved. Then labour hours can be deduced. On the other hand, one will have to be careful not to write the contract on a dayworks basis, as the employer has little control over labour hours spent.

Another way would be to write the schedules by inserting rates e.g. for labour, and getting tenderers to fill in amount e.g. hours. The final tender cost would be the basis for adjudicating, and payments are based on the contractor's times inserted at tender stage unless the quantities change.

Inducement

The alternative way of encouraging labour use is by means of use of shadow values in tender adjudication. An example of such inducement is given below, from a labour orientated contract.

A contract can be orientated to provide employment (labour intensive) and also to provide opportunity for training and advancement for employed people, which is a more long term benefit to independent states. To meet the first requirement the tenderer is required to state the components of labour and mechanization he intends using, and the labour usage indicated must be maintained during the contract. Failing this, a deduction could be made as follows: The tendered labour usage will be accounted for by subtracting $3 (the shadow value) multiplied by the number of labour unit-days indicated, from the tender price for tender adjudication only.

To meet the training requirement, the contractor is required to appoint on a full or part-time basis human resource officers to perform the following duties:

(i)   Recruit suitable labour

(ii)   Allocate labourers to appropriate tasks

(iii)   Ensure they are trained in their tasks

(iv)   Arrange promotion whenever this is possible, to more senior tasks

(v)   Appoint suitable local sub-contractors to perform selected tasks (e.g. road construction and maintenance, gabion work) and provide training where necessary. (By local is meant wholly owned by citizens)

(vi)   Assist local sub-contractors to establish, administer, manage and expand. For the latter purposes assistance with obtaining plant and vehicles will be taken into account.

In adjudicating tenders, value should be attached to sub-contracts awarded to local contractors by subtracting say 10% of the value of such sub-contracts awarded from the tender price. If the contractor fails to arrange the indicated sub-contracts the same amount may be withheld from payment.

The cost of the training officer is to be indicated by the tenderer but if he fails to perform the duties indicated to the satisfaction of the engineer his costs may be withheld. His cost shall be subtracted for purposes of adjudicating tenders.

The use of labour-intensive methods by private contractors employed on work in the public sector is a relatively new development. Several methods have been used by the public sector to encourage contractors to adopt labour-intensive methods. By monitoring work-in-progress the most successful can be identified. At this stage it is suggested that the public sector consider initiating training programmes to augment those now being run by contractors in order to augment the supply of suitably trained personel.

MANAGEMENT CONTRACTS

The procedure of awarding management contracts, as done in the U.S.A. for various reasons, has great potential in developing areas. The procedure is to select an experienced and acceptable contractor to award sub contracts to constructors. The Management Contractor is responsible for running the job, but more importantly in developing areas, for transfering technology to local contractors. He should be required in the contract

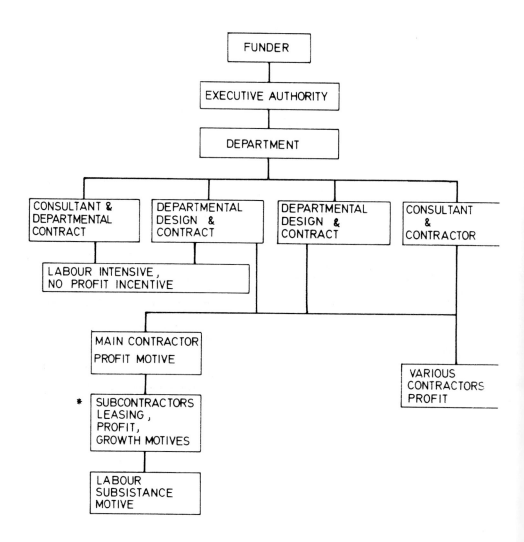

* THIS IS THE ONLY POSSIBILITY FOR LOCAL DEVELOPMENT

Fig. 13.1 Hierarchy diagram of construction using labour methods

documents, to provide formal and site training, financial, plant and labour management, and to finance fledgling subcontractors. Project costs may be higher initially, but as local contractors gain experience they may start to undercut by emphasizing labour methods rather than plant, and the local economy should benefit owing to increasing skills.

REFERENCES

Deepak Lal, 1978. Men or Machines. Geneva. I L O.

de Veen, J J, 1980. The Rural Access Roads Programme Appropriate Technology in Kenya. ILO, Geneva.

IBRD (International Bank for Reconstruction and Development), 1971. Study of the Substitution of Labour and Equipment in Road Construction, Phase I: Final Report Washington IBRD.

IBRD, 1974. Study of Substitution of Labour and Equipment in Road Construction, Phase II: Final Report Washington, IBRD.

IBRD, Coukis, B. et al. 1983. Labour-based Construction Programs A Practical Guide of Planning and Management. Oxford: OUP for the World Bank.

IBRD, 1986. A Study of the substitution of Labor and Equipment in Civil Construction. A Research and Implementation Project, Project Completion Report, Washington IBRD.

I L O, 1988. Guide to Tools and Equipment, Mainwaring and Webster.

McCutcheon, R.T., 1983. District Pilot Project at labour-intensive road construction and maintenance: Botswana. Final Report. CTP 17. Geneva, ILO.

McCutcheon, R.T. 1988. Labour-intensive road construction and maintenance: the implications for South AFrica of other sub-Saharan experience. Joint SAICE-ATC Convention, Pretoria.

McCutcheon, R.T. and Stephenson, D., 1988. Labour intensive methods and private enterprise. Joint SAICE-ATC Convention, Pretoria.

Stephenson, D. and Paling, W.A.J., 1987. Planning model for water resources development in a developing country. Water Systems Research Group, University of the Witwatersrand.

CHAPTER 14

ENVIRONMENTAL AND SOCIAL IMPACT ASSESSMENT

INTRODUCTION

The effects of construction of water resources projects on a basin and its environment needs careful assessment. It is standard practice to evaluate the impact of dam construction on river systems, and guidelines for this have been issued by U.S. agencies e.g. AID, the World Bank, and a sub-committee of the International Commission on Large Dams (ICOLD).

The U.S. Water Resources Council attempted to achieve a balanced assessment by requiring account of effects on four factors (Petersen, 1984):

National economic development
Environmental quality
Regional economic development
Social effects

While not deprecating environmental effects, project construction in developing areas is likely to have more social and economic effects, and to be desperately required in many cases. The local population thus tends to attach less importance to environmental factors than in developed countries. Short term aesthetic impacts may be of almost no local concern, but guidance is needed where long term impacts result, e.g. soil erosion or deforestation. On the other hand, social impacts can be enormous and warrent careful evaluation. In fact, it is quite likely environmental effects usually will be positive as economic and social improvement will facilitate better farming practices and a more conscientious population. The most glaring environmental problem in many developing basins is soil degradation and erosion which must be tackled with positive agricultural and reclamation policies. Without such policies, further deterioration can render more tracts of land useless and result in sedimentation of rivers and reservoirs on an increasing scale (Goodman, 1984).

Environment quality is generally associated with cultural, ecological and aesthestic properties. Culture can include lifestyle, history, sociological and economic aspects. Ecology can include floral, fauna, water, geophysics and climate. Aesthetics embraces the senses, including visual and tourism potential.

A matrix type analysis is often used done over four stages, namely:

i) Preliminary identification and data identification.

ii) Preliminary appraisal to identify likely environmental effects.

iii) Identification for and inventory of data.

iv) Assessment of effects.

The effect of both construction activities and the presence of a reservoir need assessing. The effects of the following actions were suggested by ICOLD for study:

Construction

Water use

Presence of reservoir

Operation

Socio-economic activity

The following effects have to be considered:

Socio-economic impact

Geophysical impact

Impact on water

Impact on flora

Impact on fauna

ENVIRONMENTAL IMPACTS OF PROPOSED WATER RESOURCES DEVELOPMENT

The now standard environmental impact assessment format of first world countries is not necessarily the way to assess impacts in developing areas because priorities may differ in a developing country. The desire for development may be very important at this stage, giving environmental issues a lower priority. Social and economic impacts may be so important that aesthetic conservation and pollution may receive less attention and expenditure.

On the other hand, some environmental aspects, particularly those with long lasting and economic consequences, may receive a high priority. Soil erosion is such an example. Another is the effect of river control on flood plains used for agriculture. Others are the relocation of people, and the intrusion of development into a relatively stable but undeveloped society. Whether development is a desirable feature may be debateable by some first world conservationists, but to the people who have no opportunity now, development may be all important because it may be accompanied by education, jobs and money.

Cada and Zadraga (1981) presented much of the following considerations in their report 'Environmental Issues and Site Selection Criteria for Small Hydro Projects in Developing Countries'.

Impacts on Water and Related Resources:

Physical impacts:

Streamflow patterns will change with the construction of dams. Water will be released through turbines over certain hours of the day, and the river will therefore be dry at other times for a number of kilometres downstream to where other inflow occurs, except for pools. Minimum low flow releases may be required during non-generating periods, reducing flows available for power production.

Flood flows will be less frequent and only those overtopping the spillway will be passed downstream. Therefore, inundation of flood plains downstream will be less frequent and river bank agriculture may be reduced.

During filling of a large reservoir (which may require a year or more), downstream flows would be reduced to minimum low-flow requirements except for releases through turbines.

River diversion during construction will only affect a limited length of river bed but flow rates will be essentially unchanged.

Evaporation from reservoirs could decrease the mean annual river yield, especially in arid and semiarid areas.

Groundwater recharge is unlikely to be affected at dam sites as solid rock exists beneath reservoir walls and basins in many cases.

Chemical impacts:

The quality of water from upstream could be important as it may contain fertilizers and insecticides that could affect the water quality in reservoirs.

Reservoir stratification by chemicals or temperature should be monitored but at small hydro power developments is not normally expected to be significant as the water level will fluctuate a lot. Multiple level outlets can be used to average outflow quality so reservoir quality will not deteriorate appreciably over time and so that any downstream water temperature criteria can be met.

Contaminants are expected to be low in concentration causing little adverse conditions to whatever fish or biological life there is.

Biological impact:

Changes in flow or load could change river bed conditions and adversely affect fish breeding or aquatic vegetation.

Migratory fish could be blocked by construction of a dam wall. The reservoir may be stocked with fish and these could escape down the penstocks unless the screens are adequate.

The type of fish in the reservoir may be different from that in the river owing to the more steady water level and quality, and different water temperature. Stocking of the reservoir should be considered, particularly if the local people could fish.

Lagoons and tidal mouths sometimes exist where life could be affected.

Dredging and in-stream excavation during construction are likely to affect water turbidity and consequently fish life downstream.

The construction of a reservoir could result in the deposit of vegetation and trees in the basin but these do not form a noticeable load.

Sedimentation impact

The biggest impact of a large reservoir could be on the suspended load in the river downstream of the reservoir. Over 90% of silt (primarily the coarser particles) could be deposited in the reservoir so that the water downstream will be deficient in sediment. There may therefore be potential for degradation of the river bed and erosion of the banks. The frequency and amount of deposition of sediment on the river banks downstream will decrease. High suspended sediment loads can adversely affect operation and maintenance of the turbines.

Impacts on Land and Related Resources

Physical impacts:

The reduction of silt deposits was discussed above.

Upstream of a dam there will be no change unless agricultural patterns change. It is only likely to be dry-land farming or pumped irrigation which could affect the catchment.

If irrigation or some form of intensive farming does occur in the catchment, soil erosion could increase unless soil conservation measures are used.

Arable land upstream of a dam could be inundated. This land may be farmed for subsistance purposes or used for grazing, and the farmers should be compensated, for example, by offering pumped water supplied to higher lying land.

Settlements could be inundated and a survey of the water-line should be made to establish if isolated buildings will be affected.

Roads across the river may exist and the dam wall may have to be built to accommodate a re-aligned road. Service roads will be required to reach settlements cut off.

Archeological resources may be inundated by the reservoir.

Biological impacts:

Site clearing will destroy vegetation. Grazing is included in arrable land above.

Wild birds or fauna could be affected by construction.

Existing migration routes could be affected.

The reservoir shore line will often be more accessible than the prior river bed for livestock or game to water at.

Wetlands or nature areas downstream could be affected.

Sediment deposits in the reservoir headwaters during floods may be arable during non flood periods.

Impacts on Public Health

The reservoir may be a breeding ground for mosquitoes and other insects.

Interbasin transfer of water could introduce new diseases e.g. bilharzia.

Influx of construction workers may result in spread of disease e.g. AIDS.

The construction camp and completed reservoir will result in a net immigration as well as tourist points. Sanitary facilities and water supplies will be required for these people.

Socio-Economic Impacts

The injection of money by way of construction and provision of a water supply and other facilities at the project sites will serve as a node for growth. Migration from outlying areas can be expected, and concentrations of people will make it easier to provide water supply and sanitation, transport and health and educational facilities.

Being a convenient artery, the reservoir may attract tourism, fishing and possibly commercial and recreational boating.

ASSESSMENT OF ENVIRONMENTAL IMPACT ON DAMS

ICOLD (International Committee on Large Dams, 1982, 1985) has initiated considerable research into the impact of dams and the assessment of such impacts. It has published a number of booklets assisting planners of dams and other major water works. Apart from technical references such as the matrix method description (ICOLD, 1982) there are guides on what to look for in making assessment. ICOLD (1985) has through careful research and documentation provided a wealth of information from dams in temperate, tropical, subtropical, and severe winter regions. The chapters on tropical, semi tropical and arid regions are probably of most relevance to developing countries.

Although the committee's report concentrates on natural aspects (e.g. water, land, fauna, flora, climate) there are sections on economic and social aspects, and on man. However, these are primarily from a first world point of view. The fact that many dams are designed by first world engineers for third world countries is a problem very few recognise unless they understand the way of life of the third world population. It is not necessary to preserve the rural way of life, but to see how those people can best adopt to the new lifestyles and facilities thrust upon them, i.e. don't treat the people as curios, but try to understand what they need and want.

Matrix Method

ICOLD (1982) proposed a matrix method of analysing environmental impact of dams. The proposed matrix is set up in terms of actions and facilities causing impacts (rows) and effects (columns). The rows are given in Fig. 14.1. The columns are below

Economics;

    Industrialisation and commercialisation

    Employment

    Tourism

    Crop and livestock farming

    Communications

    Trade – local finance

    Re-assessment of land value

Society;

   Social acceptance

   Recreation

   Local landmarks and character

   Appearance

   Domestic water supply

   Land acquisition

   Diminishing rural population

   Protection against natural dangers

   Health

   (Other impact on man)

Geophysical Impact;

   Morphology

   Erosion

   Suspended load

   Bed load

   Aggradation, sedimentation

   Slope stability

   Induced earthquakes

   Soil salinity

   Flooding

   New hydromorphic ground

   Reclamation and drainage

   Tidal changes

   (Other)

Impact on Water;

   Biology

   Physics and chemistry

   Salinity

   Solid loads, turbidity

   Temperature

   Evaporation

   River flow

   Water loss

   Water table

   (Other)

280

Climate;

New mesoclimate

Impact on Terrestrial Flora and Aquatic Flora;

Forest

Moor and fallow

Grass growth

Cropped land

Higher plants

Active microflora

Phytoplanton

Rare/endangered plants

(Other flora)

Terrestrial and Aquatic Fauna;

Mammals

Birds

Insects

Reptiles and amphibians

Economic fish species

Other fish species

Macro-invertebrates

Zooplankton

Microorganisms

Rare/endangered species

(Other fauna)

Instructions for filling in the matrix:

The matrix proposed by the Committee on the Effects of Damming and the Environment is intended to provide a means of listing and evaluating, even if in only qualitative or relative terms, the impact of individual dams and related construction work on specific parts of the environment. It is hoped it will enable designers and decision makers to take steps to control detrimental effects and accentuate beneficial ones.

The matrix derives from the one drawn up by the US Geological Survey. It takes the form of a table in which the rows deal with the effects on the economic, social, geophysical, hydrological, climatic and biotic environment and the columns detail the characteristics of action involved, with distinctions between the use for which the water is destined, the type of action, the zone concerned, physical corrective action and institutional action.

This splits up the overall impact of the dam into a series of readily grasped unit impacts, in which each factor and action are clearly distinguished.

The completed matrix must always be accompanied by a written commentary and justify the user's interpretation.

Completing the matrix for a given dam can be broken down into six steps, as follows:

1. Scan list A to find all the major actions involved in the project, beginning with those comprising the basic purpose of the development (A10) which are numbered in decreasing order in the Water Uses column. After this come the other actions specific to the dam and its construction (A20).

2. From list E, take all the environmental factors liable to be affected by these actions.
   The matrix now works as a check list to make a systematic inventory of impacts.

3. The areas affected by the impacts are then marked (A30)

4. Each impact is evaluated with symbols which introduce the concepts of relative importance, degree of certainty, duration and delayed effects. Any one effect may be evaluated in several different ways, depending on the time at which, and place where it occurs. The symbols Y or N denote whether or not, respectively, the effect was purposely sought in the design.
   The symbols can thus express the dynamic progression of the effects, but it is stressed that not all the symbols need be used for a successful matrix, nor does the use of the symbols dispense with the need for a written commentary.

5. Arrows may be used to relate effects to one or more actions to illustrate "chain reactions" or how two or more effects may be interdependent; the arrows reveal the feedback loops.

6. The last step is to imagine what corrective measures might be suitable to remedy or mitigate the effects listed and determine if they in turn might not cause harm to the ecosystem.

The foregoing shows that while the matrix can be used merely as an inventory, its full potential is realised by reading it both vertically and horizontally, to understand the dynamic aspects of the whole system.

But however it is used, it must not be forgotten that the matrix is above all an aid to thinking. It cannot replace a thorough ecological survey by specialists in the field. On the contrary, it must convince users of the need to seek expert advice on each specific point.

Feedback loops, examples of corrective measures, and test applications are also given in ICOLD (1982), making the publication a worthwhile reference.

Munn (1975) goes back to basic philosophy in suggesting procedures for environmental impact assessments. Administrative procedures, objectives and alternative methods are discussed.

REFERENCES

Cada and Zadraga, 1981. Environmental Issues and site Selection Criteria for Small Hydro Projects in Developing Countries. Oak Ridge Natl. Lab. Tennesse.

Goodman, A.S., 1984. Principles of Water Resources Planning, Prentice Hall, N.J., 563p.

ICOLD (International Committee on Large Dams), 1982. Dams and the Environment, Bull 35, Paris.

ICOLD (International Committee on Large Dams), 1985. Dams and the Environment, notes on regional influences.Bull. 50. Paris.

Munn, R.E. (Ed.) 1975. Environmental Impact assessment, Principles and Procedures. Scope 5, 2nd Edn. John Wiley & Sons.

Petersen, M.S., 1984. Water Resources Planning and Development. Prentice Hall.

CHARACTERISTICS OF DAM AND RESERVOIR

NAME OF DAM

Name of River _____

Country _____

Owner _____

Purpose(s) of project _____

DAM

type _____

height (m) _____

crest length (m) _____

RESERVOIR

catchment area (km²) _____

average annual runoff (m³/s) _____

average precipitation (mm) _____

evaporation (mm) _____

maximum recorded flood (m³/s) _____

spillway capacity (m³/s) _____

reservoir storage volume (m³) _____

reservoir length (km) _____

area of reservoir :

  max level _____

  min level _____

perimeter length (km) _____

REMARKS

ENVIRONMENT

geology _____

relief _____

climate _____

population _____

vegetation and crops _____

wild life and _____

domestic animals _____

position in catchment :

  . head reach _____

  . downstream _____

  . estuary/delta _____

river width (m) _____

grade at sight(%) _____

water quality _____

The matrix columns (SOCIAL AND ECONOMIC IMPACT) include:
E 101 INDUSTRIALISATION / COMMERCIALISATION, E 102 EMPLOYMENT, E 103 TOURISM, E 104 CROP AND LIVESTOCK FARMING, E 105 COMMUNICATIONS, E 106 TRADE LOCAL FINANCE, E 107 RE-ASSESSMENT OF LAND VALUE

A — WATER USES
- A 101 IRRIGATION
- A 102 ENERGY
- A 103 DRINKING WATER
- A 104 RIVER REGULATION
- A 105 INDUSTRIAL USE
- A 106 NAVIGATION
- A 107 FIRE SUPPLY
- A 108 FISHING
- A 109 WATER SPORTS
- A 110

PHYSICAL FACTORS
- A 201 PRESENCE OF DAM
- A 202 RESERVOIR
- A 203 DIVERSION OF WATER
- A 204 CONSTRUCTION SITE
- A 205 COFFERDAMS
- A 206 BUILDINGS
- A 207 DEFORESTATION
- A 208 QUARRIES AND BORROW PITS
- A 209 SHIPPING LOCK
- A 210 PENSTOCK/POWER TUNNEL
- A 211 RELEASE OF WATER
- A 212 INTAKES/PERMANENT DIVERSION
- A 213 TRANSMISSION LINES
- A 214

AREAS AFFECTED
- A 301 SUBMERGED AREAS
- A 302 RESERVOIR SURROUNDINGS
- A 303 RESERVOIR FLUCTUATION ZONE
- A 304 RIVER ABOVE RESERVOIR
- A 305 RIVER BELOW RESERVOIR
- A 306 IRRIGATION CANALS
- A 307 GROUND WATER
- A 308 SEA COAST
- A 309

CORRECTIVE ACTIONS
- A 401 FISH MANAGEMENT RESTOCKING
- A 402 GUARANTEED RIVER FLOW
- A 403 TOURIST DEVELOPMENT
- A 404 WATER LEVEL CONTROL
- A 405 INFRASTRUCTURE
- A 406 REAFFORESTRATION
- A 407 EROSION CONTROL
- A 408 DREDGING
- A 409 RELEASE THROUGH DAM
- A 410 AUXILIARY DAM
- A 411 COMPENSATION RESERVOIR
- A 412 FLOATING BOOM
- A 413 WATER CATCHMENT CONTROL
- A 414 TREATMENT OF WATER
- A 415 AMELIORATIVE INDUSTRIES
- A 416 RESETTLEMENT
- A 417
- A 418
- A 419

LAWS
- A 501 TAXES AND CHARGES
- A 502 RE-ASSESSMENT OF LAND VALUE
- A 503 TOWN AND COUNTRY PLANNING
- A 601

Fig. 14.1 Sample Matrix (ICOLD, 1982)

EXAMPLE OF USE OF SYMBOLS

FOR USE WITH FIG. 14.1

WATER USES (A10)

I   First
II   Second
III Third
etc . . . .

IMPACT OF PROJECTS

+ beneficial
− detrimental
x possible but difficult to quantify
   without specific studies

IMPORTANCE	1	minor
	2	moderate
	3	major

PROBABILITY	c	certain
	p	probable
	i	improbable
	n	not known

| DURATION | T | temporary |
| | P | permanent |

WHEN	I	immediate
	M	medium term
	L	long term

| EFFECT DESIGNED FOR | Y | yes |
| | N | no |

AUTHOR INDEX

SUBJECT INDEX